Transport Mechanisms of Tryptophan in Blood Cells, Nerve Cells, and at the Blood-Brain Barrier

Journal of Neural Transmission
Supplementum 15

Transport Mechanisms of Tryptophan in Blood Cells, Nerve Cells, and at the Blood-Brain Barrier

Proceedings of the International Symposium, Prilly/Lausanne, Switzerland, July 6-7, 1978

Edited by P. Baumann

Springer-Verlag Wien GmbH

Dr. rer. nat. Dipl.-Chem. Pierre Baumann
Head of the Laboratory of Biochemistry
Centre de recherche psychopathologique
Clinique psychiatrique universitaire de Lausanne
Switzerland

Originally published by Springer-Verlag/ Wien in 1979
Softcover reprint of the hardcover 1st edition 1979

With 54 Figures

Library of Congress Cataloging in Publication Data. Main entry under title: Transport mechanisms of tryptophan in blood cells, nerve cells, and at the blood-brain barrier. (Journal of neural transmission : Supplementum ; 15) 1. Tryptophan metabolism-Congresses. 2. Biological transport-Congresses. 3. Blood cells-Congresses. 4. Neurons-Congresses. 5. Blood-brain barrier-Congresses. I. Baumann, Pierre, 1944— . II. Series. QP562.T7T72. 612'.398. 79-10471

ISBN 978-3-7091-2245-7 ISBN 978-3-7091-2243-3 (eBook)
DOI 10.1007/978-3-7091-2243-3

Preface

At the previous meeting of the International Study Group for Tryptophan Research (ISTRY) at Madison, in 1977, two main problems arose in the discussions about the serotonin hypothesis of depression and schizophrenia: first, the enzymatic conversion of tryptophan into serotonin, second and predominantly, its availability to the brain. The latter problem was the topic of the symposium held at the Psychiatric Clinic of the University of Lausanne, of which the proceedings are presented in this volume. Indeed, the results obtained with tryptophan as an antidepressant or hypnotic are conflicting. Also, the hypothesis of impaired transport of tryptophan from blood to brain in mental diseases, remains to be confirmed.

This symposium has given some answers to questions concerning nutritional factors, the influence of competing amino acids and the relevance of the binding of tryptophan to blood albumin—all necessary parameters regulating the uptake of tryptophan into the brain.

For human studies, a peripheral model was considered useful. For this reason, the transport systems in the intestine, in blood cells, at the blood-brain barrier, and in the nervous system were compared.

One of the aims of the symposium was also to fill the gap between applied and basic sciences in that it brought together biological psychiatrists and biochemists, familiar with the very molecular basis of the transport systems. Thanks to the convergence of representatives of these disciplines, the meeting had not only scientific value, but was also an event with a particularly friendly "ambiance".

I am very grateful to Prof. Sourkes, who, in Madison, encouraged me to organize this meeting and to Prof. Wurtman, who from his great experience, gave many helpful suggestions. We owe to him and to Springer-Verlag the publication of these proceedings. I am indebted to Prof. Müller, who, at the head of a clinic, orientated mainly towards psychotherapy, urged me to organize a symposium ranging from pure biochemistry to biological psychiatry.

My thanks are extended especially to Mrs. Jacqueline Bourquin, assisted effectively by Mrs. Battaglia, Mrs. Marin and Miss Perey, whose work for several months was marked by selfless devotion in making the meeting possible.

Prilly/Lausanne, February 1979 **P. Baumann**

Welcome

Ladies and gentlemen, I would like first of all to welcome you to this hospital. Some of you have covered a long distance to come here and you all wish to discuss the question of tryptophan. I hope this meeting will be profitable and interesting and the discussion between researchers will be fruitful. As a psychiatrist I have always been fascinated by the idea that one day correlations may be established between emotions, or what the ancients called the passions, and cerebral metabolism. Some psychiatrists may be somewhat reluctant to consider this perspective, particularly if they are oriented in the same way as I am, that is, in the direction of psychotherapeutics and psychoanalysis. But if we are to deserve our reputation of men of science, we must be prepared to explore every aspect of the human being and to accept the results of our research, whether or not we may like them, whether or not they may correspond to our individual philosophy. At this university clinic which I have been proud to direct for nearly 20 years, the tradition has always been one of eclectism as far as research is concerned, one of being open to all hypotheses. My predecessor, Professor Steck, had already been one of the pioneers of biological treatments and had carried out experiments on the application of insulin to psychotics, long before Sackel. We have the privilege of being here at a cross-roads between two cultures, between the French-speaking and the German-speaking countries. I ardently wish that, at this same place, we may be at a cross-roads between a psychobiological research and a psychodynamics research.

I have expressed my conviction that you will have an interesting scientific discussion. But I also hope that you will feel at home along the shores of Lake Leman, in a country that men such as Rousseau and Voltaire have loved, this country where the good wine and the fondue are much appreciated. Be welcome and please accept my thanks for having answered our call.

Ch. Müller

The organisation of this symposium was made possible by the generous financial support of:

Fonds National Suisse de la Recherche Scientifique

CIBA-Geigy S.A., Basle

Société Coopérative Migros Vaud, Lausanne

Nestlé S.A., Vevey

Fabriques de Tabac Réunies S.A., Lausanne

Galenica Représentations, Berne

CILAG-CHEMIE S.A., Schaffhouse
Dépt. Janssen Pharmaceutica

Wander S.A., Berne

Banque Cantonale Vaudoise, Lausanne

Crédit Foncier Vaudois, Lausanne

Rhodia-Pharm S.A., Genève

Contents

Journal of Neural Transmission, Suppl. 15, 1—13 (1979)

Interaction Between Parallel Transport Systems Examined With Tryptophan and Related Amino Acids*

H. N. Christensen and **M. E. Handlogten**

Department of Biological Chemistry, The University of Michigan, Ann Arbor, Michigan, U.S.A.

With 6 Figures

Summary

Most of the neutral amino acids are transported across the plasma membrane by two or more parallel transport systems. In this study, we have limited transport interactions to System *A* and *L* by studying amino acids not transported by System *ASC*. Results are presented to emphasize that such amino acids as leucine, phenylalanine and tryptophan do enter the Ehrlich cell to substantial degrees by System *A*. We show how the phenomenon of competitive stimulation presents a strong argument that these systems operate between the same two compartments, extracellular and cellular. To discover what is needed to bring two amino acids into the transport relation shown by tryptophan and methionine, we arbitrarily set up two classes of amino acids: (1) those whose steady-state gradients will be increased when System *L* is deleted; (2) those whose gradients will instead be decreased. On blockading System *L* with 2-aminonorbornane-2-carboxylic acid applied in symmetry to the two sides of the plasma membrane, we show as predicted that the gradient maintained for methionine is strongly increased, that for tryptophan decreased. The initial rate of uptake of all the amino acids mentioned is highly sensitive to the nature of the cellular amino acid pool. A high influx by exchange, relative to net influx, appears to characterize the amino acids that respond to competitive stimulation. Albumin-bound tryptophan appears not directly accessible to the tested carriers.

---

* Support is acknowledged from the Institute of Child Health and Human Development, Grant HD01233, National Institutes of Health, U.S.P.H.S.

Introduction

The System A-System L Dichotomy

My remarks will offer some historic orientation as to the division of the transport of tryptophan into cells among transport systems. There are some aspects of this division of transport among systems that have been misinterpreted, or which will bear reexamination on other grounds. I will leave it for those who follow me to consider how well the expectations arising from the work I'll report are met for the tissues of the brain.

The central point to be made is that almost every amino acid moves in and out of almost every cell of the animal organism by two or more parallel but distinct transport systems. It had come to be appreciated as early as 1947 that different transport systems serve for the anionic, the cationic and the neutral amino acids, at least for renal tubular transport, and as early as 1961 (*Johnstone* and *Scholefield*) that more than one system was required even to account for the uptake of the neutral amino acids. The subsequent discovery that each neutral amino acid divides its transport into and out of the cell among two or more systems was based on several kinds of evidence (*Akedo* and *Christensen*, 1962; *Oxender* and *Christensen*, 1963; *Christensen et al.*, 1965; *Inui* and *Christensen*, 1966; *Christensen* and *Handlogten*, 1968). The evidence of Fig. 1 will serve, however, to show the necessity for that conclusion. Here we see (upper left) that little or no correlation exists between the activity of each of 30-odd amino acids in inhibiting the transport of leucine and that of glycine into the Ehrlich cell. At the lower right we see that the activity of these numerous amino acids in inhibiting the transport of leucine is rather well correlated with their ability to inhibit phenylalanine transport. This behavior is what we should expect if leucine and phenylalanine enter the cell mainly by a common route. The correlation, lower left, between the effectiveness of amino acids in inhibiting the uptake of alanine and that of glycine is also correlated, although distinctly less well. This somewhat poorer correlation is well understood now, because we recognize that a third amino acid system *(ASC)* participates in the transport of alanine and several other amino acids, although not in the transport of glycine (*Christensen et al.*, 1967). The system traced out by the plot at the lower right is now recognized as the Na^+-independent one. In some other tissues a Na^+-dependent system nearly specific to glycine is also present, and also one specific to β and γ (or only γ) amino acids.

The two systems predominating in the behavior of Fig. 1 we gave the arbitrary designations *A* and *L*. These systems appear to be nearly

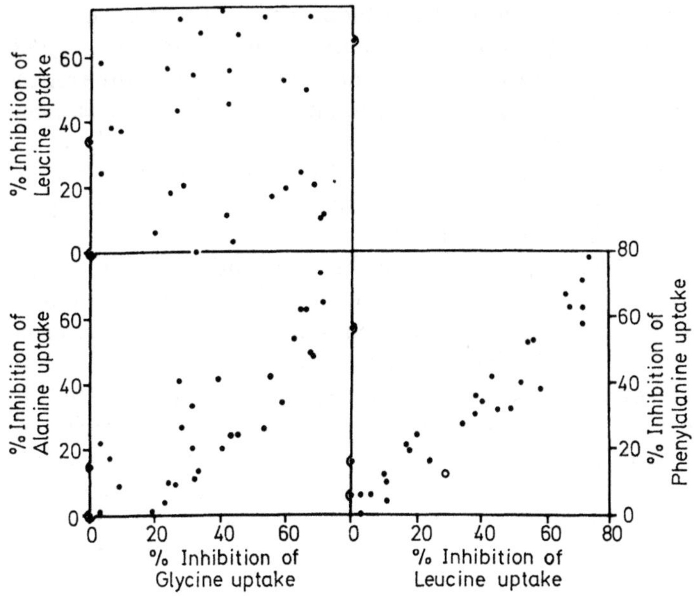

Fig. 1. To what degree is the effectiveness of each of 36 amino acids as inhibitors of transport of one amino acid correlated with their effectiveness as transport of another amino acid? Constructed from data of *Oxender* and *Christensen* (1963). See text for discussion. Reproduced with permission from *Christensen,* Perspectives in Biology and Medicine *10,* 471 (1967)

ubiquitous, not only in cells in their usual organismal context, but also in these cells grown in isolated culture. The pattern appears to be similar for nervous tissues, as shown by *Blasberg* and *Lajtha* (1966), but no doubt we will hear much about those tissues. Because of the temptation to regard the red blood cell as a model cell for experimental study, I should remind you that in the mammal the Na^+-dependent systems of this cell largely disappear on maturation. Furthermore, even in the nucleated and reticulated red blood cells, System *A* appears to be absent, its place apparently taken by the glycine system. This absence leaves the quite different Na^+-dependent system *ASC* as dominant for the transport of alanine, serine, threonine, hydroxyproline, and the like. Presumably because this *ASC* system does not tolerate substitution of Li^+ for Na^+, amino-acid-dependent Li^+ uptake was discovered in the Ehrlich cell (*Christensen* and *Handlogten,* 1978) rather than in the erythrocyte. In the mature red cell of normal sheep an otherwise similar system was found Na^+-independent (*Young* and *Ellory,* 1977). Perhaps in that case the Na^+-dependency rather than the system is lost on maturation.

System *L* of the nucleated red blood cell of the pigeon shows only a weak exchanging property (*Eavenson* and *Christensen*, 1968). These various results show that we must expect some differences in the properties of the transport systems in their various occurrences. Another example is revealed by *α*, *α*-dicyclopropylglycine, which appears to serve as a substrate for System *L* in rat brain slices, although the molecule is too bulky for mediated transport in the Ehrlich cell (*Christensen*, 1972).

An important point that has been misunderstood about the *A* and *L* dichotomy is illustrated in Fig. 2, *left* for leucine, and *right* for tryptophan. The behavior of phenylalanine was similar to that shown here for leucine. Probably because we chose the letter *L* to designate what at first we called the *leucine-preferring* system, the substantial role played by System *A* in the transport of these three and related amino acids is often overlooked. We now discourage the term *leucine-preferring*, also *alanine-preferring*, and treat the letters *A* and *L* as no more than arbitrary designations. The second and third curves in each section of Fig. 2 show how much the initial rate of uptake of the test amino acid is decreased by eliminating the contribution of System *A*, either by adding a specific inhibitory analog, 2-(methylamino)-isobutyric acid (AIB), or by omitting Na^+ from the medium. The

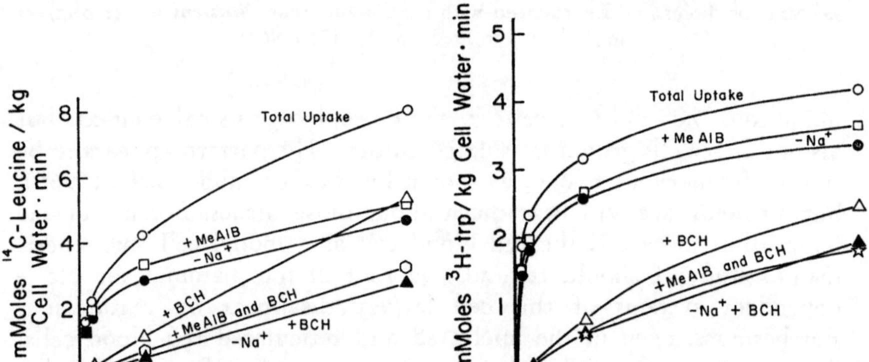

Fig. 2. Division of the uptake of leucine (left) and tryptophan (right) into the Ehrlich cells among transport systems. Uptake observed during 30 sec at 37 °C from Krebs-Ringer bicarbonate medium at pH 7.4. Two methods have been used to suppress uptake by System *A*: Addition of 2-(methylamino)-isobutyric acid at 20 mM; replacement of Na^+ in the medium with choline. The norbornane amino acid (BCH) has been used at 10 mM to suppress uptake by System *L*. When both contributions are suppressed at once, the residual influx corresponds to the non-saturable rate, which should not however, be assumed to represent simple diffusion

differences between the top curve and the fourth curve show the somewhat larger contributions of System *L*.

When 4 % albumin was added to 32 μM tryptophan in HEPES-buffered Ringer solution, 86 % of the tryptophan was removed from reactivity for transport into the Ehrlich cell. This measurement is based on the increase of the apparent K_m for uptake from 32 to 230 μM. This result indicates that the albumin-bound tryptophan is unreactive for transport into this cell. The same proved the case for the rat hepatocyte in primary culture. Albumin had no effect on uptake of the norbornane amino acid.

The Phenomenon of "Competitive Stimulation"

I want now to consider a peculiarity observed for tryptophan uptake by the Ehrlich cell (*Jacquez*, 1963). The simultaneous addition of methionine leads in a minute or two, to a stimulation of tryptophan uptake rather than the expected inhibition. The origin of this phenomenon, sometimes called *competitive stimulation*, was briefly a matter of controversy. Because of a widespread reluctance to accept the *A* and *L* dichotomy, *Guroff et al.* (1964) and *Jacquez* and *Sherman* (1965) argued that the *stimulated* amino acid might preferentially form its complex with the transport carrier by replacing the already bound *stimulating* amino acid, rather than by direct combination. In that way stimulation could presumably occur within a single transport system. *Oxender* and I (1963) showed how the phenomenon observed by *Jacquez* and *Guroff* was a predictable consequence of (a) the rapid cellular accumulation of methionine via System *A* and (b) the exchange of that internalized methionine for external tryptophan via System *L*.

Jacquez provided in 1967 an additional piece of evidence that persuaded him to embrace our explanation. Although the uptake of tryptophan ordinarily is mainly Na^+-independent (Fig. 2, *right*), the stimulation of its uptake by methionine requires the presence of Na^+. Fig. 3 shows that the increase in tryptophan uptake was always less than the simultaneous decrease in methionine uptake, as our theory requires. As the observation time is shortened, stimulation disappears, also as the theory requires (data not shown).

Let me point out that this relation between the uptake of tryptophan and methionine (seen also for a few other amino acid pairs) provides very important evidence that Systems *A* and *L* are operating between the same two compartments, extracellular and cellular. Even if we must inevitably take into account some compartmentation of the cell interior (see *McKee et al.*, 1978), there appears

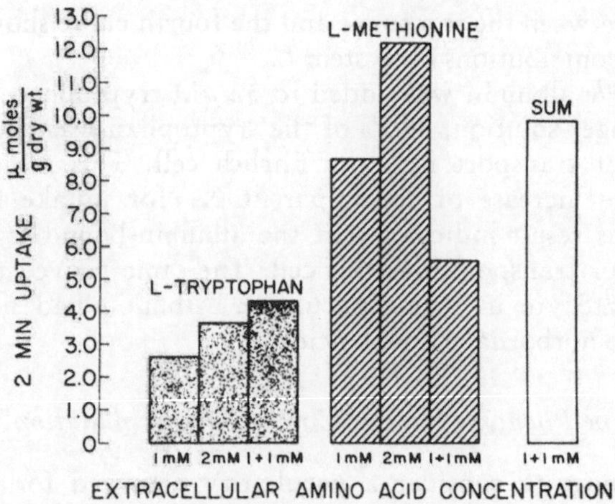

Fig. 3. When tryptophan and methionine are presented together, the increase in tryptophan uptake is more than compensated by the decrease in methionine uptake. We may therefore speak of methionine as the *inhibited* amino acid, and tryptophan as the *stimulated* amino acid. Reproduced with permission from *Schafer* and *Jacquez,* Biochim. Biophys. Acta *135,* 741 (1967)

to be no evidence that one of these transport systems feeds its substrate directly into protein synthesis, or into a separate anatomical compartment. The behavior shows that the uptake of methionine by System *A* of the Ehrlich cell presents to System *L* a sufficient methionine gradient to drive an accelerated tryptophan uptake. If System *A* placed methionine in a compartment separate from the one causing its exodus by System *L,* this behavior would not be seen.

Why do tryptophan and methionine behave in this way? The interaction between two amino acids when they are presented to the cell together, so that one is the *stimulator* and the other the *stimulated,* can be expected to depend on relations between their kinetic properties for transport, and also on their concentrations. So far methionine, *DL*-p-fluorophenylalanine and leucine have all been shown to stimulate tryptophan uptake, usually at 1 mM concentrations of each. The intensity of their action decreases in the order listed. Several other combinations failed to show consistent stimulation (*Schafer* and *Jacquez,* 1967). We have now found cycloleucine mildly stimulating to tryptophan uptake, and methionine strongly stimulating to uptake of the norbornane amino acid, weakly to phenylalanine, leucine, and tyrosine uptake. The pairs, leucine and tryptophan, norleucine and tryptophan, leucine and tyrosine, and leucine and the norbornane amino acid showed only mutual inhibition in our hands. A further finding that cycloleucine stimulates a large increase in uptake of the norbornane amino acid shows that metabolizability of the amino acids concerned is not needed for a product of System *A* transport to become immediately reactant in System *L* within the cell.

The property that makes methionine effective, shared with cycloleucine, appears to be a fairly even division of uptake between Systems *A* and *L*. What then is the essential property of tryptophan uptake? Slowness of uptake in the absence of competitive stimulation is not the answer. Fig. 2, *right*, and similar experiments place the K_m at less than 0.03 mM and V_{max} at 2.8 mmoles per kg cell water · min. These correspond to a rate quotient, V_{max}/K_m, of about 100 min^{-1}, a high rate for System *L*.

Fig. 4 shows that tryptophan is highly sensitive as to its initial rate of uptake, to manipulation of the internal amino acid pool; but so are phenylalanine and the norbornane amino acid (BCH). The four curves of each of the 3 sections of Fig. 4 show the initial rate of uptake of these three amino acids after manipulations of the endogenous amino acid content of Ehrlich cells as described in the legend. The ratio of the maximum velocities shown by the methionine-loaded cells to those shown by the AIB-treated cells in Fig. 4 (the norbornane amino acid 9.4, tryptophan 8.8, phenylalanine 6.1) take the same order as the intensities of competitive stimulation due to methionine (the norbornane amino acid 63 %, tryptophan 51 %, phenylalanine 13 %; 2 min at 1 mM levels). So do the ratios for V_{max}/K_m for cells treated by the two methods. Because exchange has been minimized, influx into the AIB cells will be rich in *net* entry by System *L*. The ratios of System *L* to System *A* influx (cf. Fig. 2) also fall in that order, whereas the rates of entry by System *A* take the inverse order, phenylalanine > tryptophan > norbornane amino acid (immeasurably slow). These results show that total *net* entry via the two systems must take the latter sequence. A high influx by exchange, relative to *net* influx, appears therefore to determine competitive stimulation.

Nature of the Steady-State

I have developed elsewhere the view that the consequence of discharge by the two systems into the same cellular compartment should produce net uptake by the stronger System *A*, and net exodus by the reversal of System *L*; also that this net exodus by reversal of System *L* should conserve a substantial part of the energy originally invested in accumulation by System *A* (*Christensen*, 1976 a, b). This behavior is indeed shown by methionine, since there can be no question that energy is conserved when a part of the accumulated methionine is exchanged for external tryptophan.

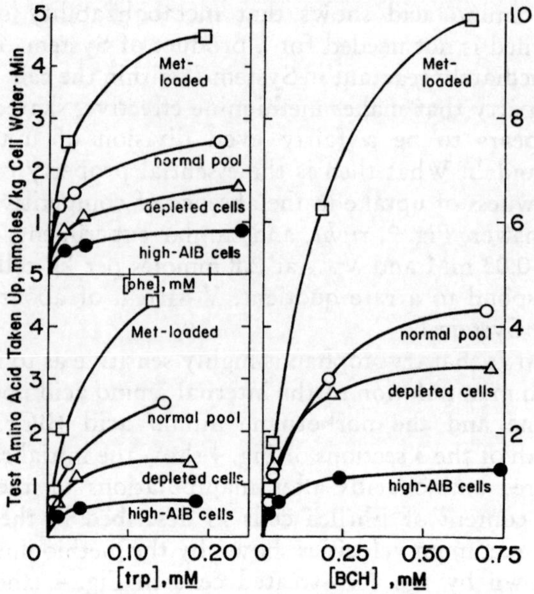

Fig. 4. Comparison of sensitivity of initial rates of uptake of three amino acids to modification of the cellular pool of the amino acids in the Ehrlich cell. The figure shows only the uptake inhibitable by the norbornane amino acid (BCH), *i.e.* uptake by System *L*, observed during 30 sec at 37 °C from Krebs-Ringer bicarbonate medium at pH 7.4. The upper curve in each case applies for cells previously incubated 5 min in 2 mM methionine; the second curve, cells not intentionally manipulated as to amino acid content; the third curve, cells depleted of amino acids by incubation in hypotonic medium at $[Na^+] = 30$ mM, $[K^+] = 35$ mM; and the fourth curve, uptake by the so-called AIB cells in which AIB replaces about 70 % of the endogenous amino acids (*Christensen et al.*, 1973, 1974). The amino acid analyzer was used to see how much these treatments modified the cellular amino acid pool in the present experiments. Simple depletion eliminated three-quarters of the endogenous amino acids (60 % of 6 large-moleculed, 77 % of 7 small amino acids). Methionine treatment eliminated 45 % of 6 large, 39 % of 7 small, replacing them with methionine equal to 43 % of the initial pool. AIB treatment eliminated 81 % of the 6 large, 82 % of 7 small, replacing them with 35 % AIB

Fig. 5 shows my 1962 picture of this behavior. In the lower diagram an amino acid *M* is shown transferring energy supplied by way of System *A*, so as to accumulate another amino acid *L* via System *L*, called unwisely in 1962 the *exchanger*. The upper diagram shows the entering amino acid (let us consider the designation *G* arbitrary) exitting by reverse operation of System *L*. It is this exodus which we now expect to cause the generation of ATP or whatever high-energy compound participates in driving the forward operation of System *L* (*Garcia-Sancho et al.*, 1975), given that a sufficient concentration of an external amino acid such as tryptophan is not supplied.

An energized system operating in reverse would appear to provide an ideal arrangement for restraining exodus, a purpose for which we

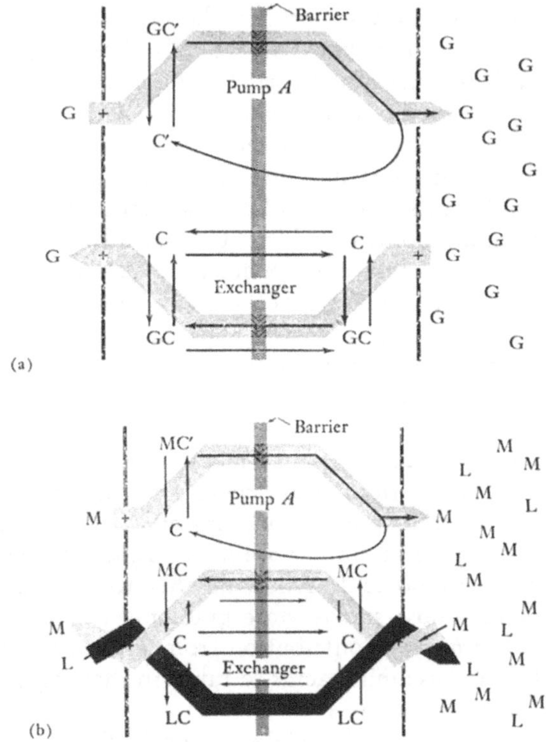

Fig. 5. Our 1962 visualization of the ways in which two distinct but parallel transport systems could operate for the entry and the exodus of an amino acid. The amino acid *G* or *M* can be accumulated by System *A* operating alone, to generate a gradient steeper than can be maintained by the weaker System *L*. Therefore a steady state is set up in which net uptake occurs by System *A*, and (upper diagram) net exodus occurs by System *L*. The latter system is here given the name *exchanger*, which is objectionable because System *L* can produce net transport as well as exchange. In the lower diagram another amino acid *L* is also provided in the extracellular environment, this amino acid being transported mainly by System *L*. The figure originally served to show how energy invested in pumping by System *A* could serve to concentrate other amino acids into the cell by exchange through System *L*. The figure serves here, however, to pose 2 forms of a more recent question: If no amino acid *L* is provided in the environment (upper figure), will energy be conserved upon the reversal of System *L*, to synthesize ATP or some other energy-supplying metabolite? In the physiological context, to what degree is energy conserved alternatively in these two ways? Reproduced with permission from Biological Transport, 1st ed. New York: W. A. Benjamin. 1962

could hardly trust simple leakage. Restraint of exodus is of manifest importance in the maintenance of the nutrition of cells of the intestinal mucosa in the post-absorptive state. The distribution of the Na^+-dependent and Na^+-independent pumps between the brush-border and the basolateral poles of epithelial cells favors that model. Such restraint is also likely to serve in the regulation of nutrient and metabolite distribution for almost all the cells of the higher animal.

Turning back to questions raised by the case of tryptophan and methionine, we can predict that various amino acids will behave quite differently in the interaction between Systems *A* and *L*, now to be referred to their flux contributions at the steady state. Using our results with the Ehrlich cell to guide us, we can arbitrarily designate as Type I certain amino acids which at the steady state will come to enter mainly by System *A* and to depart by System *L*, and whose gradients would therefore be steepened if System *L* were deleted. We can understand that the gradients of such amino acids as methionine (Type I in my provisional terminology) should be steepened by the selective deletion of System *L*. What would then be the consequence of the deletion of System *L* for the amino acids which enter by exchange with the accumulating methionine? Under conditions producing competitive stimulation, these gradients would obviously be decreased by that deletion, since System *L* becomes, under the influence of methionine or similar amino acids, the principal route of net entry for these amino acids. Allow me then to designate as Type II those amino acids whose steady-state gradients will be decreased by selectively preventing any contribution from System *L*. Obviously we may expect the various amino acids to range in their properties all the way from our stereotypes I to II.

But how can we selectively delete System *L*? By careful manipulation in the experiment of Fig. 6 we have applied the norbornane amino acid at both inward and outward surfaces of the Ehrlich cell membrane at concentrations balanced so that little or no net *trans* effect is produced in either direction. This figure shows that under these conditions the loss of previously accumulated methionine is not modified by externally applied tryptophan, whereas if no BCH is added at all, methionine moves outward and tryptophan inward. As the figure shows, the consequence is that a much higher steady-state gradient of methionine was maintained when System *L* was prevented from contributing to methionine migration. At the same time the uptake of tryptophan was sharply reduced by the blocking of System *L* in this experiment. When phenylalanine was used in similar experiments in place of methionine, we were unable to show intensification of its steady-state accumulation by the balanced presence

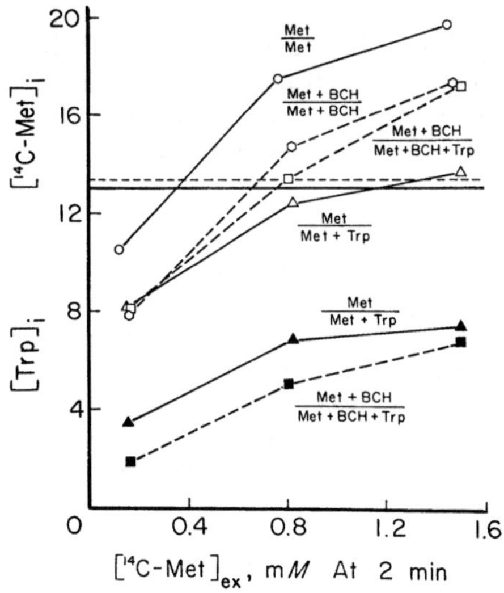

Fig. 6. Effect of external tryptophan on the level of external methionine required to maintain a previously established apparent cellular methionine level of 13.1 to 13.4 mM (horizontal lines), when all transport systems are available (solid lines), also when System *L* is blocked with the norbornane amino acid applied in balance inside and outside (dashed lines). The Ehrlich cells had first been depleted of endogenous amino acids by 3 successive incubations, 5 min each, in forty volumes of low-Na^+ Krebs-Ringer bicarbonate medium (266 mOsmolar; $[Na^+] = 6$ mM). Portions of the cells after centrifugation were then reloaded by 5-min incubation in (a) 2 mM ^{14}C-methionine or (b) 3 mM ^{14}C-methionine + 10 mM ^{3}H-norbornane amino acid, or in a corresponding amino-acid-free Krebs-Ringer bicarbonate medium. The washed cells were then incubated for 2 min in 115 volumes of a Krebs-Ringer bicarbonate medium containing 0.1, 0.8 or 1.5 mM ^{14}C-methionine, with or without the same concentration of tryptophan. The level of the norbornane amino acid was thus brought to an apparent level of 29 mM in the cell water. The final incubation was terminated as usual by adding 3 volumes of ice-cold Krebs-Ringer bicarbonate medium in which choline replaced Na^+. Centrifugation for 2.5 min was followed by scintillation counting to determine the two labeled amino acids, and cellular tryptophan was determined with Ehrlich's reagent. The upper slanted lines show the apparent internal methionine levels obtained after 2 min at the 3 different external levels. The point at which each of these lines crosses the corresponding horizontal line representing the initial internal methionine level measures the distribution ratio at which neither exodus nor uptake of methionine has occurred. The above described locations of the added amino acids are summarized on the plot by placing in the numerator of a fraction the abbreviations for the amino acids present internally, and in the denominator for those present externally at zero time. BCH = the norbornane amino acid, 2-aminobicycloheptyl-2-carboxylic acid. The distance between the lower lines show that, simultaneous with abolition of its action on the net flux of methionine, tryptophan uptake has been decreased by 46, 24, and 9 %, respectively, at the 3 tryptophan levels

of the norbornane amino acid, as measured by the technique of Fig. 6 for methionine.

We are looking for a better way to delete System *L* than by symmetrically overloading it, for example, by suicidal binding to a chloramine derivative of the model substrate for that system. Synthetic work in this direction is underway.

References

Akedo, H., Christensen, H. N.: Transfer of amino acids across the intestine: a new model amino acid. J. Biol. Chem. *237*, 113—117 (1962).

Blasberg, R., Lajtha, A.: Heterogeneity of the mediated transport systems of amino acid uptake in brain. Brain Res. *1*, 86—104 (1966).

Christensen, H. N.: Biological Transport, 1st ed., p. 60. Reading, Mass.: W. A. Benjamin. 1962.

Christensen, H. N.: Some transport lessons taught by the organic solute. Persp. in Biol. and Med. *10*, 471—494 (1967).

Christensen, H. N.: Nature and roles of receptor sites for amino acid transport. In: Advances in Biochemical Psychopharmacology, Vol. 4 (*Costa, E., Greengard, P.*, eds.), pp. 39—62. New York: Raven Press. 1972.

Christensen, H. N.: Metabolite transport at cell membranes. In: Transport Phenomena in the Nervous System (*Levi, G., Battistin, L., Lajtha, A.*, eds.), pp. 3—12. New York: Plenum. 1976 a.

Christensen, H. N.: On the routes of renal tubular resorption of amino acids. In: Proc. 6th Int. Congr. Nephrol. (*Giovannetti, S., Bonomini, V., D'Amico, G.*, eds.), pp. 134—138. Basel: Karger. 1976 b.

Christensen, H. N., Handlogten, M. E.: Modes of mediated exodus of amino acids from the Ehrlich ascites tumor cell. J. Biol. Chem. *243*, 5428—5438 (1968).

Christensen, H. N., Handlogten, M. E.: Cellular uptake of lithium via amino acid transport System *A*. Biochim. Biophys. Acta *512*, 598—602 (1978).

Christensen, H. N., Oxender, D. L., Liang, M., Vatz, K. A.: The use of N-methylation to direct the route of mediated transport of amino acids. J. Biol. Chem. *240*, 3609—3616 (1965).

Christensen, H. N., Liang, M., Archer, E. G.: A distinct Na^+-requiring transport system for alanine, serine, cysteine, and similar amino acids. J. Biol. Chem. *242*, 5237—5246 (1967).

Christensen, H. N., de Cespedes, C., Handlogten, M. E., Ronquist, G.: Energization of amino acid transport studied for the Ehrlich ascites tumor cell. Biochim. Biophys. Acta *300*, 487—522 (1973).

Christensen, H. N., de Cespedes, C., Handlogten, M. E., Ronquist, G.: Modified transport substrates as probes for intramembrane gradients. Ann. N.Y. Acad. Sci. *227*, 335—379 (1974).

Eavenson, E., Christensen, H. N.: Transport systems for neutral amino acids in the pigeon erythrocyte. J. Biol. Chem. *242*, 5386—5396 (1967).

Garcia-Sancho, J., Sanchez, A., Handlogten, M. E., Christensen, H. N.: Unexpected additional mode of energization of amino-acid transport into Ehrlich cells. Proc. Nat. Acad. Sci. U.S. *74*, 1488—1491 (1975).

Guroff, G., Fanning, G. R., Chirigos, M. A.: Stimulation of aromatic amino acid transport by p-fluorophenylalanine in the Sarcoma 37 cell. J. Cellular Comp. Physiol. *63*, 323—331 (1964).

Inui, Y., Christensen, H. N.: Discrimination of single transport systems: The Na^+-sensitive transport of neutral amino acids in the Ehrlich cell. J. Gen. Physiol. *50*, 203—224 (1966).

Jacquez, J. A.: Carrier-amino acid stoichiometry in amino acid transport in Ehrlich ascites tumor cells. Biochim. Biophys. Acta *71*, 15—33 (1963).

Jacquez, J. A.: Competitive stimulation: Further evidence for two carriers in the transport of neutral amino acids. Biochim. Biophys. Acta *135*, 751—755 (1967).

Jacquez, J. A., Sherman, J. H.: The effect of metabolic inhibitors of transport and exchange of amino acids in Ehrlich ascites cells. Biochim. Biophys. Acta *109*, 128—141 (1965). (Page 140 should be examined.)

Johnstone, R. M., Scholefield, P. G.: Factors controlling the uptake and retention of methionine and ethionine by Ehrlich ascites carcinoma cell. J. Biol. Chem. *236*, 1419—1424 (1961).

McKee, E. E., Cheung, J. Y., Rannels, D. E., Morgan, H. E.: Measurement of the rate of protein synthesis and compartmentation of heart phenylalanine. J. Biol. Chem. *253*, 1030—1040 (1978).

Oxender, D. L., Christensen, H. N.: Distinct mediating systems of neutral amino acids by the Ehrlich cell. J. Biol. Chem. *238*, 3686—3699 (1963).

Schafer, J., Jacquez, J. A.: Transport of amino acids in Ehrlich ascites tumor cells: Competitive stimulation. Biochim. Biophys. Acta *135*, 741—750 (1967).

Young, J. D., Ellory, J. C.: Substrate specificity of amino acid transport in sheep erythrocytes. Biochem. J. (London) *162*, 33—38 (1977).

Authors' address: Dr. *H. N. Christensen*, Department of Biological Chemistry, The University of Michigan, Ann Arbor, MI 48109, U.S.A.

Journal of Neural Transmission, Suppl. 15, 15—27 (1979)

A Comparison of Amino Acid Transport and Ouabain Binding in Brain Endothelium and Salivary Epithelium Studied *in vivo* by Rapid Paired-Tracer Dilution

D. L. Yudilevich, F. V. Sepúlveda*, J. C. Bustamante, and G. E. Mann**

Department of Physiology, Queen Elizabeth College, University of London, London, U.K.

With 6 Figures

Summary

Amino acid transport at the luminal side of brain capillary endothelium and the basal side of salivary epithelium were compared using a non-destructive, first-circulation, paired-tracer dilution method. In the brain, the reference molecule was an intravascular marker (Crone's method) whereas in the salivary gland the reference was an extracellular marker of similar size to the test molecule. The unidirectional flux was related to a maximal uptake, U:

$$U = 1 - \text{test concentration/reference concentration}$$

Uptake and cross-inhibition experiments in brain suggest the presence of only a long-chain neutral L-amino acid transport system (leucine, phenylalanine, tryptophan, tyrosine, valine, methionine). In contrast, results in the salivary gland suggest 4 transport systems: 1. large neutral, 2. small neutral (alanine, serine), 3. basic (lysine) and 4. acidic (aspartic, glutamic). The same method was applied to localize binding sites. Ouabain bound very significantly to the salivary epithelium but not at all to the brain endothelium. The method described is of interest since it can be extended to any organ and possibly to man.

---

* Present address: ARC Institute of Animal Physiology, Babraham, Cambridge, CB2 4AT, U.K.

** World University Service Studentship.

Introduction

The transport across the blood-brain barrier of sugars (*Crone*, 1965), some amino acids (*Yudilevich, de Rose*, and *Sepúlveda*, 1972) and several other metabolic substrates has been shown to be mediated by specific carrier systems (see review by *Pardridge* and *Oldendorf*, 1977) which seem to be located in the capillary endothelium (*Goldstein, Csejtey* and *Diamond*, 1977). These mechanisms appear to be absent in other capillaries. For example, in heart (*Yudilevich* and *Martín de Julián*, 1965; *Durán* and *Yudilevich*, 1974; *Rose* and *Goresky*, 1977) and in skeletal muscle (*Crone* and *Thompson*, 1973) the carriers which transport molecules into the tissue appear to exist in the parenchymal cells and not in the endothelial cells.

This description, referred to as serial barriers to blood tissue exchange (*Yudilevich*, 1970), was the basis of *Martín de Julián* and *Yudilevich*'s (1964) theoretical analysis to quantify capillary permeability, extravascular spaces and to assess cellular uptake from the simultaneous dilution of an intravascular reference and an extravascular test tracer. More recently, cellular uptake has been measured avoiding elaborate mathematical analysis (*Yudilevich*, 1976). In this approach the reference tracer is an extracellular marker of a size comparable to that of the test tracer. The initial, or maximal, cellular uptake estimated in this way can be treated as a unidirectional flux across the basolateral membranes of parenchymal cells, in the case of epithelial tissues, in a manner analogous to that originally used by *Crone* (1965) in studies of transport across brain capillaries.

In this paper we extend the paired tracer method to the study of amino acid transport and ouabain binding in the salivary gland. These findings are compared with similar studies carried out in the blood-brain barrier, some of which have been reported (*Yudilevich et al.*, 1972). Four transport systems seem to mediate the uptake of amino acids at the basal side of the salivary epithelium whereas only one of these systems is present in brain. Ouabain binding was only observed in the salivary gland.

Methods

Dog Brain

Detailed experimental procedures have previously been published (*Crone*, 1965; *Yudilevich* and *de Rose*, 1971; *Yudilevich et al.*, 1972). In the anaesthetized animal a rapid intracarotid injection (0.7 ml in 1 sec) of a solution containing the isotope mixture was followed immediately by collection of the venous effluent from the sagittal sinus. The injectates were saline solutions, containing L-(^{3}H) or L-(^{14}C)amino acids and ^{22}Na, prepared either without or with unlabelled L-amino acid. The injectates for the studies with (^{3}H)ouabain contained ^{22}Na or (^{14}C)sucrose as the extracellular reference molecule. In the same dog, up to eight successive indicator dilution experiments could be performed (*Yudilevich* and *Sepúlveda*, 1976).

Cat Submandibular Salivary Gland

Experiments were performed on anaesthetized animals (35 mg/kg i.p. Nembutal). A close-arterial preparation was established by ligating all branches of the external and common carotid arteries except the submandibular and lingual arteries (*Darke* and *Smaje*, 1972). The natural blood supply to the gland was maintained via the submandibular artery whereas the lingual artery was cannulated retrogradely for the injection of isotopes. All tributaries draining into the external jugular vein were tied, except for the branch draining the gland. The external jugular vein was cannulated and blood flow recorded using a drop recorder.

Following a 1—3-second isotope injection (100 μl) into the lingual artery of resting glands the venous effluent was collected into 15 successive vials (2 drops of blood into each). The injectates were saline solutions containing an extracellular and a test tracer. Initially L-(^{14}C)glucose was used as the extracellular marker for L-(^{3}H)amino acids. However, in later experiments a 10—30 % L-(^{14}C)glucose uptake relative to mannitol was observed (unpublished data) and therefore in subsequent experiments (^{14}C)mannitol was used as the reference substance. As in brain, unlabelled L-amino acids were added to the injectate to demonstrate competitive inhibition of transport. (^{14}C)sucrose was used as the reference for the (^{3}H)ouabain experiments. In the same cat, 20 to 30 successive paired-tracer dilution experiments were performed.

Results

Typical tracer dilution experiments in the dog brain and in the cat submandibular gland are illustrated in Fig. 1 a. The concentration versus time curves show that the transit times for tracer molecules in the two preparations are different: peak time is about 3 sec in brain and 30 sec in salivary gland. The percent cellular uptake of a test molecule, U, is calculated from:

$$U = \left(1 - \frac{\text{concentration of test tracer}}{\text{concentration of reference tracer}}\right) \times 100$$

It can be seen in Fig. 1 b that initial uptakes are constant in both organs but rapid backflux of the test molecule occurs in brain as shown by the decline of the L-(^{3}H)phenylalanine uptake after a certain time. Since the maximal uptake, U_{max}, is related to the unidirectional flux, an average value is computed in each case and this is the parameter used to compare the various molecules. The models describing the spaces reached by the tracers in a single circulation through brain and salivary gland are illustrated in the lower part of Fig. 1.

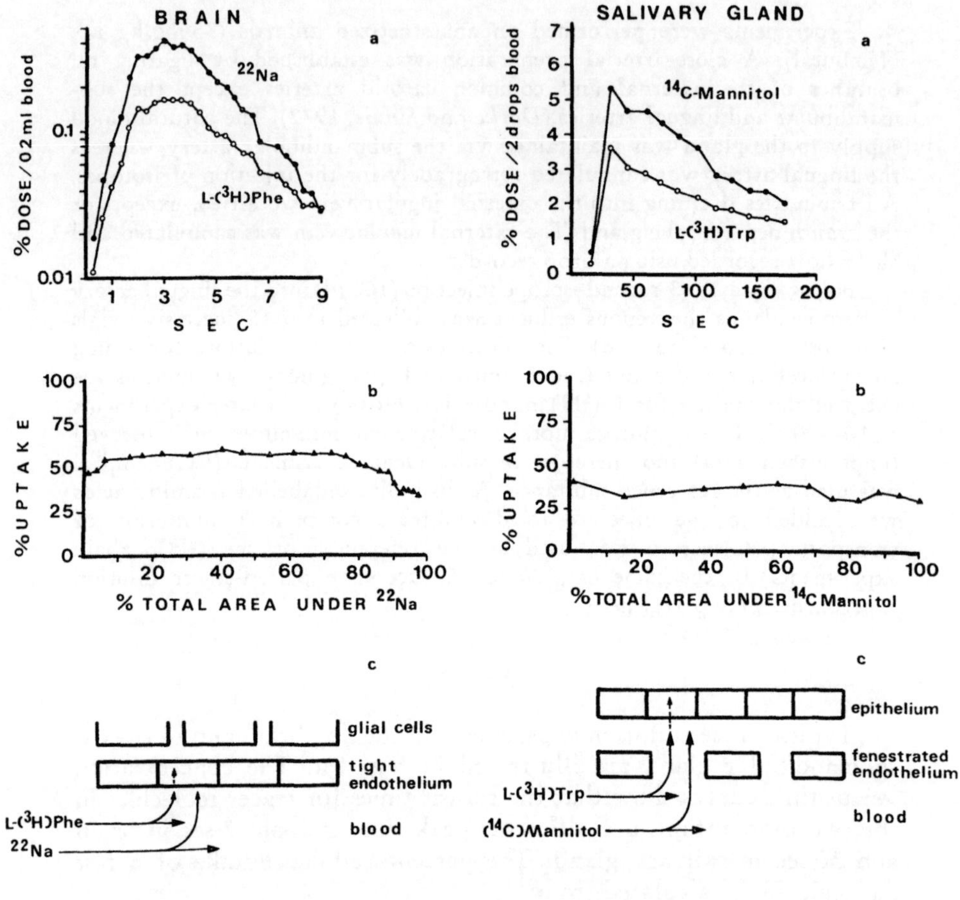

Fig. 1. Analysis of paired-tracer dilution data. In part (a) concentration-time curves are shown for ^{22}Na and L-(^{3}H)phenylalanine in dog brain and for ^{14}C-mannitol and L-(^{3}H)tryptophan in the cat submandibular gland. All tracer concentrations have been expressed as a % of the injected dose. In part (b) the % uptake of L-(^{3}H)phenylalanine and L-(^{3}H)tryptophan, calculated from 1-^{3}H/^{22}Na or 1-^{3}H/^{14}C-mannitol, has been expressed as a function of the accumulated area under the concentration-time curve of the reference tracer. Schematic models for amino acid uptake by brain endothelium and salivary gland epithelium are illustrated in part (c)

Brain Endothelial Cell Transport

Fig. 2 shows the measured uptake for all the amino acids and serotonin. The uptake was significant for some neutral L-amino acids: leucine, cycloleucine (1-aminocyclopentane-1-carboxylic acid), phenylalanine, tyrosine and valine. The uptake of histidine, which is mainly in the unprotonated form at blood pH, was also high. The uptake was not detectable for other amino acids: smaller neutral (alanine, serine, γ-amino isobutyric acid [GABA] and glycine), acidic (aspartic and glutamic) and basic (lysine and arginine). This selectivity cannot be explained in terms of the passive properties of the endothelial cell membranes or the physical characteristics of the amino acids. There was no measurable uptake of serotonin.

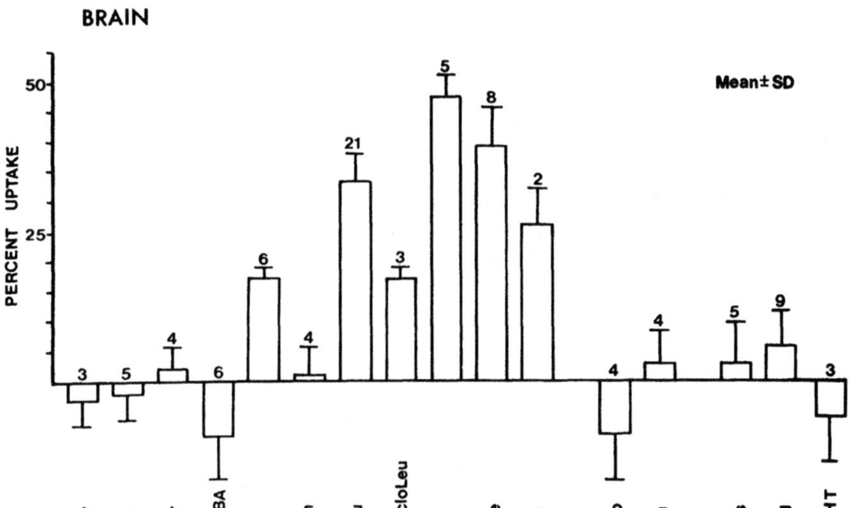

Fig. 2. Amino acid and serotonin uptake by brain endothelium. The maximal % uptake estimated for various acidic, basic and neutral amino acids and serotonin is shown; n refers to the number of observations. Standard 3-letter nomenclature is used to identify amino acids; 5-HT refers to serotonin and cycloleu to 1-amino-cyclopentane-1-carboxylic acid. These results were taken from *Yudilevich et al.* (1972) and unpublished observations

It can be shown that the uptake is a saturable process since, when the concentration in the injected solution is increased by addition of unlabelled amino acid (0.1 M), a large reduction in the tracer uptake occurs (Fig. 3). For phenylalanine and leucine self-inhibition and cross-inhibition by other transported amino acids (tryptophan, methionine and leucine) were observed whereas non-transported amino acids (serine and alanine) or D-glucose, did not reduce the uptake of the test molecule. Other inhibition experiments showed self-inhibition for histidine and the absence of cross-inhibition by the basic amino acid, arginine (*Yudilevich et al.*, 1972).

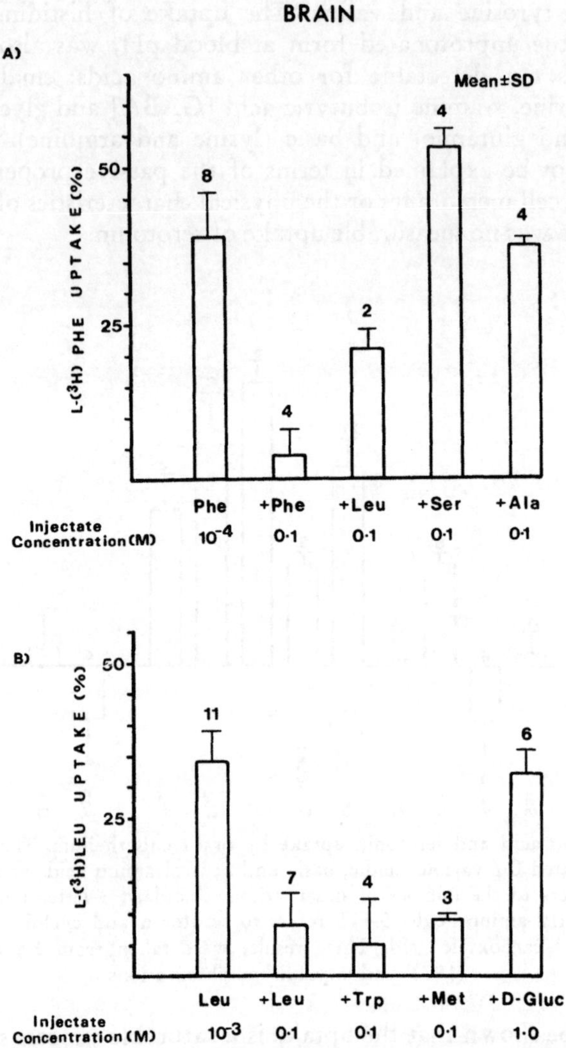

Fig. 3. Competitive inhibition of the L-system mediated transport in brain endothelium. Addition of unlabelled amino acids (phenylalanine, leucine, tryptophan and methionine) to the paired-tracer injectates at 0.1 M reduced the uptake of L-(^{3}H)phenylalanine (a) and L-(^{3}H)leucine (b). In contrast serine, alanine and D-glucose had no effect on the uptake. Data taken from *Yudilevich et al.* (1972) and unpublished results

Tryptophan seems to be part of the group of amino acids selectively transported by the endothelial cells, as it was a strong inhibitor of leucine transport (Fig. 3 b). The uptake of labelled tryptophan gave erratic results, not included in Fig. 2, which were attributed to its binding to serum albumin.

Submandibular Gland Epithelial Transport

The specificity of the amino acid transport process in the submandibular gland is shown in Fig. 4. In contrast to that observed in the brain, significant uptake was found for representatives of all the four groups of amino acids: small neutral (alanine and serine), large neutral (phenylalanine, leucine, tryptophan, methionine and histidine), acidic (aspartic and glutamic) and basic (lysine). The greatest uptakes were those for aspartic, 78 %, and glutamic, 81 %. There was no significant uptake of GABA, glycine and proline, and valine values were variable (38, 4 and 0 %).

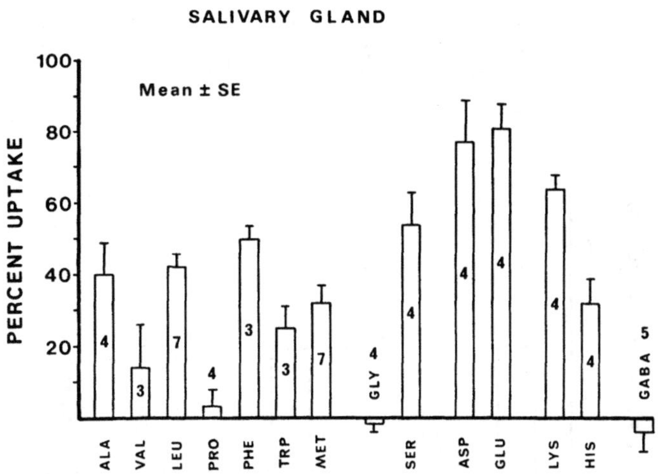

Fig. 4. Amino acid uptake by the salivary epithelium. Percent uptakes for acidic, basic and most neutral amino acids were significant. Proline, glycine and GABA were not taken up significantly. The number in each column refers to the number of observations. Part of these results have previously been reported (*Bustamante, Mann,* and *Yudilevich,* 1978)

Preliminary inhibition studies characterizing the long chain neutral amino acid carriers are illustrated in Fig. 5. In Fig. 5 a the results for labelled leucine and phenylalanine are combined since their uptakes and cross-inhibitions were similar. Methionine and tryptophan

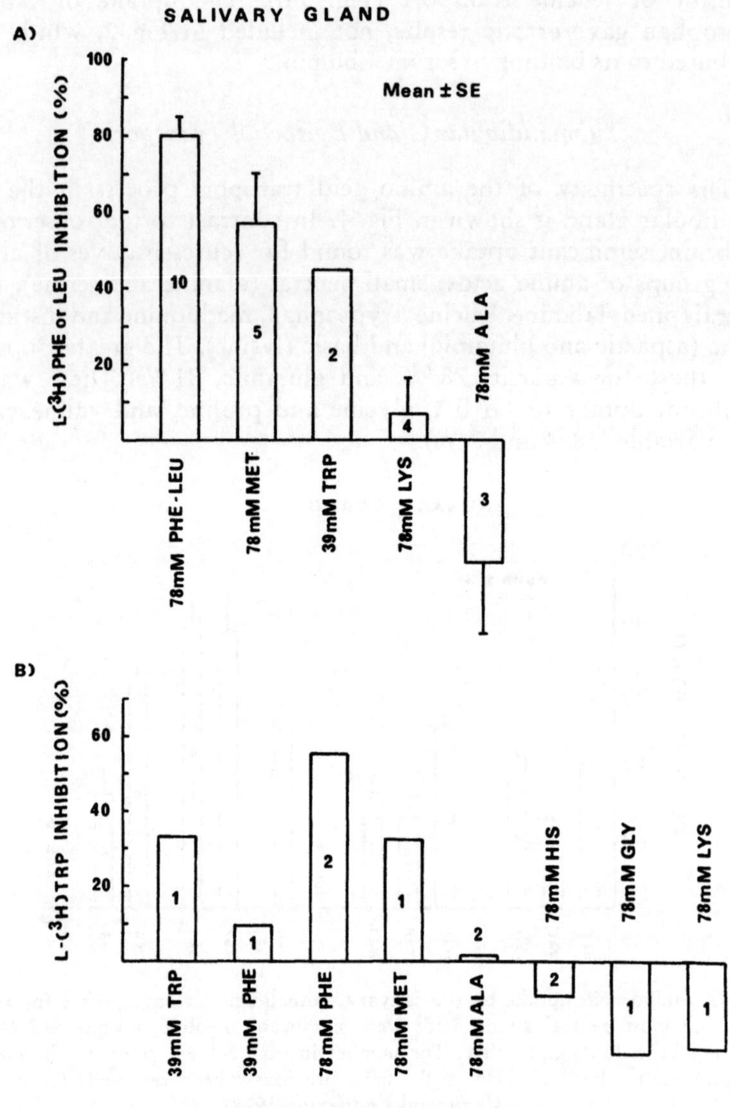

Fig. 5. Competitive inhibition of amino acid uptake in the salivary gland. Unlabelled phenylalanine, leucine, methionine or tryptophan in the injectate inhibited the uptake of L-(^{3}H)phenylalanine or L-(^{3}H)leucine (a) and L-(^{3}H)-tryptophan (b). The data for leucine and phenylalanine were combined due to the similarities in their uptake and cross-inhibitions. The number within each column refers to the number of observations

inhibited leucine and phenylalanine uptake, but the basic amino acid, lysine, and the short chain neutral, alanine, had no effect. Inhibition of tryptophan uptake is shown in Fig. 5 b. Self-inhibition and cross-inhibition by phenylalanine and methionine, but not by alanine, was observed.

Ouabain Binding

Fig. 6 shows that there was no uptake of ouabain by the brain endothelium whereas a high uptake (61 %) was found in the salivary epithelium. Ouabain injectate concentration in the salivary gland experiments was 10^{-5} M and in brain (*Yudilevich* and *Sepúlveda*, 1972) the concentrations ranged from 10^{-10} M to 10^{-6} M.

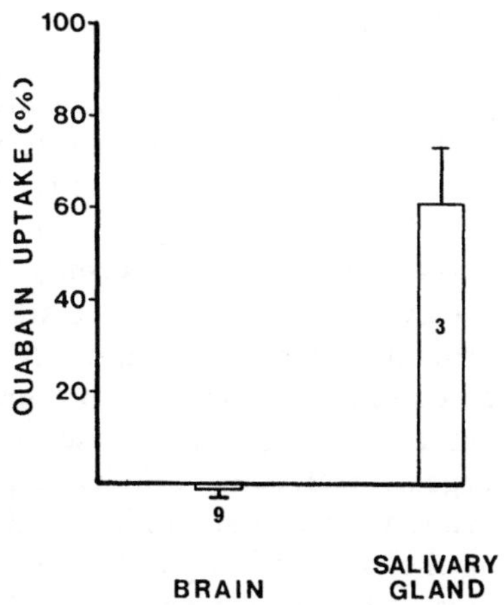

Fig. 6. Comparison of ^{3}H-ouabain uptake in the brain endothelium and salivary epithelium. Percent uptake for ouabain was calculated from $1\text{-}^{3}H/^{14}C$ as in Fig. 1 b, except that ^{14}C-sucrose was now the reference tracer. Mean ± SD values are shown, and n refers to the number of observations

Discussion

The results indicate that the capillaries of the dog brain selectively transport long-chain hydrophobic amino acids. These amino acids have been shown to be preferentially transported in other animal cells via an L-system (leucine preferring) (*Christensen*, 1969). Transport systems for the small neutral, basic and acidic amino acids seem to be absent from the brain capillaries of the dog. Another example of only a "leucine preferring" carrier has recently been described in the colon of the new-born pig which, in contrast with the L-system originally described by *Christensen*, is strongly sodium-dependent (*Sepúlveda* and *Smith*, 1978).

Studies on blood-brain transfer of amino acids in the rat suggest that, as in our study, the neutral amino acids are handled by a single L-type transport system (*Oldendorf*, 1971) and further studies confirm the absence of the A-system (alanine preferring) (*Wade* and *Katzman*, 1975). However, in other reports for rat brain a high entry of alanine, serine and cysteine was found (*Baños, Daniel, Moorehouse,* and *Pratt*, 1975).

A lysine/arginine transport system has been described in the capillaries of the rat brain (*Oldendorf*, 1971; *Baños, Daniel,* and *Pratt*, 1974). This system appears to be absent in the dog, and according to recent work it may also be absent in the brain of monkey and possibly man (*Chang*, 1978).

There was no significant uptake of the putative neurotransmitter amino acids (GABA, glycine, glutamic acid) and serotonin, suggesting that they are not readily transported from blood into the brain tissue.

Accumulation of amino acids by the salivary gland duct epithelium has been inferred from their appearance in saliva during secretion (*Lindsay, Catanzaro, Ueha,* and *Hanson*, 1969). Our data constitutes the first direct evidence for specific transport systems in the basal side of the epithelium in resting glands. A whole range of L-amino acids were taken up in a single circulation through the gland, and cross-inhibition experiments suggest the presence of four carrier systems: small neutral, large neutral, acidic and basic amino acids.

The high uptake measured for a basic amino acid (lysine) by the salivary epithelium, using the same technique, supports the contention that the postulated absence of this transport system in the dog brain, is due to a real property of the tissue rather than to shortcomings in the technique.

The model proposed in Fig. 1 c for transport in the salivary gland is supported by the finding that vitamin B_{12} has a capillary permeability in accordance with its molecular weight whereas specific binding can be demonstrated at the basal side of the salivary epithelium (*Mann, Smaje,* and *Yudilevich*, 1978). Furthermore the fenestrated capillaries of the salivary gland behave as semipermeable membranes to graded-size lipid insoluble molecules (*Mann*, 1978).

It is generally accepted that ouabain binds specifically to the transport enzyme, Na^+-K^+-ATPase. The absence of ouabain uptake at the luminal side of brain endothelium was in contrast with the marked binding observed in the salivary epithelium. In agreement with our findings, light autoradiographic studies in the salivary gland have shown ouabain binding to be primarily located at the contraluminal (basal) side of the striated duct epithelium (*Bungaard, Møller*, and *Poulsen*, 1977). Our results in brain are in agreement with cytochemical localization of phosphatase activity found only at the basal surface of the cortical capillary endothelium (*Firth*, 1977). Furthermore, paired-tracer dilution of (^{14}C)sucrose and (^{3}H)ouabain in the heart revealed binding sites for ouabain in the sarcolemma which appeared to be absent in the capillary endothelium (*Durán* and *Yudilevich*, 1974).

It is interesting to note that the non-destructive paired-tracer dilution studies described in this paper could lead to similar investigations in man. Furthermore, it now appears possible to characterize transport systems at the basal side of cells in any organ.

Acknowledgements

The brain work was partly performed at the Faculty of Sciences, University of Chile, Santiago, Chile. We gratefully acknowledge a research grant from the Medical Research Council, U.K., for the work in the salivary gland. The assistance of S. Peachell and Lesley Maitland is acknowledged.

References

Baños, G., Daniel, P. M., Pratt, O. E.: Saturation of a shared mechanism which transports L-arginine and L-lysine into the brain of the living rat. J. Physiol. (Lond.) *236*, 29—41 (1974).

Baños, G., Daniel, P. M., Moorehouse, S. R., Pratt, O. E.: The requirements of the brain for some amino acids. J. Physiol. (Lond.) *246*, 539—548 (1975).

Bungaard, M., Møller, M., Poulsen, H. J.: Localization of sodium pump sites in cat salivary glands. J. Physiol. (Lond.) *273*, 339—353 (1977).

Bustamante, J. C., Mann, G. E., Yudilevich, D. L.: An *in vivo* study of amino acid carriers in the salivary epithelium. J. Physiol. (Lond.), abstract in press.

Chang, Y. F.: Lysine metabolism in the rat brain: blood-brain barrier transport, formation of pipecolic acid and human hyperpipecolatemia. J. Neurochem. *30*, 355—360 (1978).

Christensen, H. N.: Some special kinetic problems of transport. Adv. Enzymol. *32*, 1—20 (1969).

Crone, C.: Facilitated transport of glucose from blood into brain tissue. J. Physiol. (Lond.) *181*, 103—113 (1965).

Crone, C., Thompson, A. M.: Comparative studies of capillary permeability in brain and muscle. Acta Physiol. Scand. *87*, 252—260 (1973).

Darke, A. C., Smaje, L. H.: Dependence of functional vasodilatation in the cat submaxillary gland upon stimulation frequency. J. Physiol. (Lond.) *226*, 191—203 (1972).

Durán, W. N., Yudilevich, D. L.: Capillary and cellular barriers to ouabain transport in the heart. Microvasc. Res. *7*, 84—88 (1974).

Firth, J. A.: Cytochemical localization of the K^+-regulation interface between blood and brain. Experientia *33*, 1093—1094 (1977).

Goldstein, G. W., Csejtey, J., Diamond, I.: Carrier mediated glucose transport in capillaries isolated from rat brain. J. Neurochem. *28*, 725—728 (1977).

Lindsay, R. H., Catanzaro, O. L., Ueha, T., Hanson, R. W.: Site of amino acid transfer in dog parotid glands. Am. J. Physiol. *217*, 1025—1029 (1969).

Mann, G. E.: Capillary and epithelial exchange in the cat submandibular gland. Ph.D. Thesis, University of London, 1978.

Mann, G. E., Smaje, L. H., Yudilevich, D. L.: Epithelial uptake of vitamin B_{12} in the cat salivary gland. J. Physiol. (Lond.), abstract in press.

Martín de Julián, P., Yudilevich, D. L.: A theory for the quantification of transcapillary exchange by tracer dilution curves. Am. J. Physiol. *207*, 162—168 (1964).

Oldendorf, W. H.: Brain uptake of radiolabelled amino acids, amines and hexoses after arterial injection. Am. J. Physiol. *221*, 1629—1639 (1971).

Pardridge, W. M., Oldendorf, W. H.: Transport of metabolic substrates through the blood-brain barrier. J. Neurochem. *28*, 5—12 (1977).

Rose, C. P., Goresky, C. A.: Constraints on the uptake of labelled palmitate by the heart. The barriers at the capillary and sarcolemmal surfaces and the control of intracellular sequestration. Circ. Res. *41*, 534—545 (1977).

Sepúlveda, F. V., Smith, M. W.: Different mechanisms for neutral amino acid uptake by new-born pig colon. J. Physiol. (Lond.) (in press, 1978).

Yudilevich, D. L.: Serial barriers to blood-tissue transport studied by the single injection indicator diffusion technique. In: Capillary Permeability (*Crone, C., Lassen, N.*, eds.), pp. 115—129. Copenhagen: Munksgaard. 1970.

Yudilevich, D. L.: Transport across capillary and cellular barriers of the placenta studied by single circulation multiple tracer dilution. J. Physiol. (Lond.) *256*, 84—85P (1976).

Yudilevich, D. L., Martín de Julián, P.: Potassium, sodium and iodide transcapillary exchange in the dog heart. Am. J. Physiol. *208*, 959—967 (1965).

Yudilevich, D. L., de Rose, N.: Blood-brain transfer of glucose and other molecules measured by rapid indicator dilution. Am. J. Physiol. *220*, 841—846 (1971).

Yudilevich, D. L., de Rose, N., Sepúlveda, F. V.: Facilitated transport of amino acids through the blood-brain barrier of the dog studied in a single capillary circulation. Brain Res. *44*, 569—578 (1972).

Yudilevich, D. L., Sepúlveda, F. V.: The specificity of amino acid and sugar carriers in the capillaries of the dog brain studied *in vivo* by rapid indicator dilution. In: Transport Phenomena in the Nervous System (*Levi, G., Battistin, L., Lajtha, A.*, eds.), pp. 77—87. New York: Plenum Press. 1976.

Wade, L. A., Katzman, R.: Synthetic amino acids and the nature of L-DOPA transport at the blood-brain barrier. J. Neurochem. *25*, 837—842 (1975).

Authors' address: Prof. Dr. *D. L. Yudilevich*, Department of Physiology, Queen Elizabeth College, University of London, Campden Hill Road, Kensington, London W8 7AH, U.K.

Journal of Neural Transmission, Suppl. 15, 29—42 (1979)

Kinetics of Tryptophan Transport Across the Blood-Brain Barrier

O. E. Pratt

Department of Neuropathology, Institute of Psychiatry, London, U.K.

With 4 Figures

Summary

The transport of tryptophan across the blood-brain barrier *in vivo* has been analyzed kinetically to assess the influence upon it not only of the concentration of tryptophan in the blood plasma, but also of the inhibitory effect of fourteen other amino acids normally present in the circulation. The transport of tryptophan into the brain conforms to the equations of enzyme kinetics with competitive inhibition, provided appropriate modifications are made to take account of a non-saturable component (possibly passive diffusion) in the transport, and of the normal presence in the blood of other amino acids which act as competitive inhibitors. The inhibitor constant has been evaluated for each amino acid and in many cases it corresponds fairly closely with the Michaelis saturation constant for that amino acid, suggesting that tryptophan and these inhibitors share a common transport carrier. The implications of the kinetic analysis are considered for hormonal and metabolic disturbances which change the proportions of amino acids in the blood. The effect of an increase in competitive inhibition due to raised levels of amino acids is to make the tryptophan influx more directly dependent upon the level of tryptophan in the blood.

That the rate at which tryptophan is supplied to the brain is not wholly determined by the level of tryptophan in the blood was suggested by the findings of *Fernstrom* and *Wurtman* (1972) and of *Fernstrom, Larin,* and *Wurtman* (1973) who showed that the tryptophan content of brain tissue was also affected by changes in the levels of other neutral amino acids in the circulation. Raised levels of other amino acids, especially the aromatic ones, reduce considerably the rate at which tryptophan enters the brain (*Daniel, Moorhouse,* and *Pratt,* 1976; *Pratt,* 1976). Thus all work on the influx of tryptophan into the brain *in vivo,* must take account of the levels of a number of other amino acids in the blood. The complex nature of this problem has not been fully appreciated until recently and it is not yet clear how many other amino acids, which are normally present in the blood, appreciably affect tryptophan transport into the brain, or how severe are their effects.

A kinetic analysis has been carried out to assess the influence upon the influx not only of varying the concentration of tryptophan in the blood, but also of varying the concentrations of each of the other amino acids which are normally present in appreciable amounts in the circulation. It should be possible to predict, from these findings, the effects upon tryptophan influx of the changed levels of amino acids that are found in abnormal states.

To measure the influx of tryptophan into the brain (one-way movement across the cerebral capillaries) an intravenous injection of ^{14}C-L-tryptophan was given to 8—12 week old Wistar albino rats continuously at a variable rate (*Pratt,* 1974) according to a predetermined programme (*Daniel, Donaldson,* and *Pratt,* 1974, 1975, 1976; *Donaldson* and *Pratt,* 1975) so as to maintain a steady level of specific activity of radioactive labelling of the tryptophan in the blood. At the end of the injection (usually one minute) the blood was rapidly washed out of the cerebral vessels (*Daniel, Love, Moorhouse, Pratt,* and *Wilson,* 1974) and the radioactivity that had entered the various regions of the brain was measured. (This technique for measuring influx has been used in my laboratory over the last 10 years.) In separate experiments programmed intravenous injections of unlabelled tryptophan (or of various other amino acids) were given so as to maintain different raised levels in the circulation. Separate schedules of injection were determined for each amino acid. Arterial blood samples were taken both for assay of radioactivity and for estimation of concentrations of amino acids. For details of the analytical procedures used see *Daniel, Pratt,* and *Wilson* (1977 a, b, c).

The specific activity of radioactive labelling of tryptophan in the blood was calculated on the assumption that, when ^{14}C-tryptophan was injected, both free and albumin-bound tryptophan became labelled rapidly. On the other hand no prior assumption was made about whether it was the level of free tryptophan (*i.e.* that not bound to albumin) or the total concentration of tryptophan that determined its influx into the brain.

Effect on Influx of Different Concentrations of Tryptophan in the Blood

As the level of tryptophan in the blood was raised progressively, in successive experiments, the influx increased, but in a rather complex manner (Fig. 1). Although the results showed that there was saturation of a carrier system, the data could not be fitted adequately by a rectangular hyperbola passing through the origin, as is required by an equation of Michaelis-Menten type:

$$v = \frac{Vs}{s + K_t} \qquad (1)$$

in which v is the influx, s is the concentration of tryptophan in the blood plasma, V is the maximum rate of transport, and K_t is the Michaelis constant (equivalent to K_m in enzyme kinetics), *i.e.* the value of s at which v = V/2 (in the absence of inhibitors).

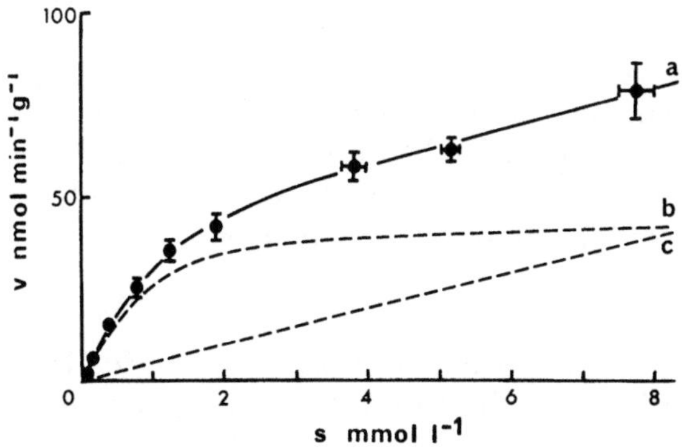

Fig. 1. The effect of increasing the concentration of free tryptophan, s, in the circulating blood upon its influx into the brain. Each point represents a separate experiment or the mean (with bars to indicate S.E.) of 3—6 experiments. The fitted line, a, is the sum of an estimated non-saturable component of transport (line c) and a saturable component (line b) according to equation (2). Data of *Daniel, Moorhouse*, and *Pratt*

A much better fit to the data was obtained by a non-rectangular hyperbola passing through the origin, as given by:

$$v = \frac{Vs}{s + K_t} + Ds \qquad (2)$$

in which the second term represents a non-saturable component of the influx (which may well be passive diffusion) and D is a first order rate constant, *i.e.* an apparent diffusion coefficient.

Estimates of the value of D were obtained in three different ways, firstly a rough estimate was obtained by fitting the best straight line by linear regression to the data for high values of s; secondly, by a direct fit of equation (2) to the data, using the method of maximum likelihood (*Mood* and *Graybill*, 1963) and, finally, by an iterative method to determine a value most consistent with Michaelis type kinetics as in equation (1) for the carrier-mediated component. In this procedure the coefficient of the non-saturable influx, D, was first assumed to have a low value and this was used to calculate the component of the influx due to the saturable process, v_t, by subtracting the estimated value of the non-saturable component, v_α:

$$v_t = v - v_\alpha . \quad (3)$$

Equation (1) was then fitted to the data corrected in this way (Fig. 1) so as to eliminate the effect of diffusion. The method of least squares was used, with statistical weighting followed by fine adjustment of the estimated values for non-linear regression (*Wilkinson*, 1961). This procedure was then repeated except that successively higher values were assumed for D. The best fit was judged to be that which gave a minimum value of the coefficient of variation of V.

These methods yielded values for D which ranged from 2.8 to 5.8, the most reliable estimate being 4.9 ± 1.4 nmole $min^{-1} g^{-1}$ of brain per mmole l^{-1} of tryptophan in the blood plasma.

Effect of Raised Concentrations of Chemically Related Amino Acids Upon the Influx of Tryptophan Into the Brain

When, the levels, in the circulating blood, of various other neutral amino acids are raised the influx of tryptophan into the brain is reduced to varying degrees, ranging from the severe inhibition caused by raised levels of the other *aromatic* amino acids, phenylalanine, tyrosine and dihydroxyphenylalanine, to the negligible effect of a raised level of 2-aminoisobutyrate (*Daniel*, *Moorhouse*, and *Pratt*, 1976, 1978 a). We showed that the influx of tryptophan into the brain conforms, at least approximately, to the kinetics of competitive inhibition of a transport carrier system (*Dixon* and *Webb*, 1964) as given by:

$$v_t = v - v_\alpha = \frac{s\,V}{s + K_t\left(1 + \dfrac{i}{K_i}\right)} . \quad (4)$$

However, a serious problem arises in attempting to estimate the values of K_t and V from data like those of Fig. 1, using equation (2), or to estimate the value of K_i (the equilibrium constant for the reaction of the inhibitor with the carrier) from data like those of Fig. 2, using equation (4), since, in work *in vivo,* when the inhibitor is an amino acid that is normally found in the blood, measurements cannot be made when i is zero.

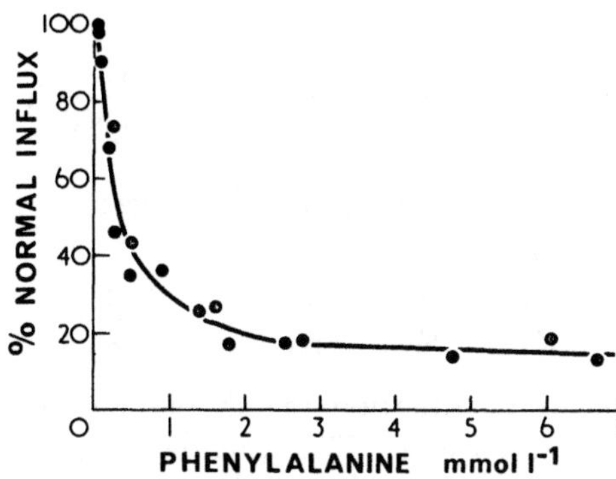

Fig. 2. The effect of raising, to successively higher levels, the concentration of an inhibitor, i (phenylalanine), in the circulation upon the influx of tryptophan into the cerebrum. Each point represents a separate experiment and a curve has been fitted, using equation (6). Data of *Daniel, Moorhouse,* and *Pratt*

The effect of this limitation is to make it difficult to estimate K_t independantly of K_i by the usual procedures of kinetic analysis (*Dixon* and *Webb,* 1964). There is a similar difficulty in estimating K_i, since in most methods it is assumed that K_t is already known. To overcome this difficulty a method based upon that of *Dixon* (1953) was used. Dixon's method had, however, to be modified to meet a further difficulty which arises because, amongst the amino acids normally found in the bloodstream, there are an unknown number of inhibitors of tryptophan transport, each having normal levels in the blood of $i_0, i_1, i_2, \ldots i_n$, and being associated with a corresponding inhibitor constant, $K_{i_0}, K_{i_1}, K_{i_2}, \ldots K_{i_n}$.

Thus a new series of terms has to be added to equation (4) to take account of the effects of these inhibitors:

$$v_t = v - v_\alpha = \frac{sV}{s + K_t\left(1 + \frac{i_0}{K_{i_0}} + \frac{i_1}{K_{i_1}} + \frac{i_2}{K_{i_2}} + \ldots + \frac{i_n}{K_{i_n}}\right)}. \quad (5)$$

When the concentration of only one inhibitor was raised from its normal level, i_x, to a new value, i, it was assumed that the concentrations of all the other inhibitors remained constant throughout the series of experiments. The new influx v_i was given by:

$$v_i - v_\alpha = v_t = \frac{sV}{s + K_t\left(1 + \frac{i_0}{K_{i_0}} + \frac{i_1}{K_{i_1}} + \ldots + \frac{i_n}{K_{i_n}}\right) + (i - i_x)\frac{K_t}{K_{i_x}}} \quad (6)$$

which can be inverted to give:

$$\frac{1}{v_i - v_\alpha} = \frac{s + K_t\left(1 + \sum_{n=o}^{n=n} I_n\right)}{sV} + \frac{K_t}{K_{i_x} sV}(i - i_x) \quad (7)$$

where $I_n = \frac{i_n}{K_{i_n}}$ and $\sum_{n=o}^{n=n} I_n$ stands for the series:

$$\frac{i_0}{K_{i_0}} + \frac{i_1}{K_{i_1}} + \frac{i_2}{K_{i_2}} + \ldots + \frac{i_n}{K_{i_n}}. \quad (8)$$

The denominator of equation (5) should be compared with the equation used by *Pardridge* (1977) to calculate the apparent Michaelis constant for transport of an amino acid into the brain, $K_m(app)_a$, in terms of the K_m values of all the amino acids which share the transport system:

$$K_m(app)_a = K_{m\alpha}\left(1 + \sum \frac{[AA]}{K_m}\right).$$

This parallel is to be expected in view of the close correspondence that we have found between the Michaelis constants and the inhibition constants for the aromatic amino acids (Table 1 and *Daniel, Moorhouse,* and *Pratt,* 1978 a).

Equation (7) was used to analyse kinetically the inhibition of tryptophan influx into the brain by raised levels of phenylalanine in the blood (Fig. 2), fitting the equation to the data by the method of least squares, with appropriate statistical weighting of the data (Fig. 3 and 4, line a). This procedure was repeated with the data from a second series of similar experiments in which the concentration of tryptophan in the blood was simultaneously raised above the normal to approximately 0.88 mmole l^{-1} (Fig. 3 and 4, line b). A third line was added representing the maximum rate of carrier-mediated influx, V, 47 ± 4 nmole $min^{-1} g^{-1}$ of brain, when the concentration of tryptophan in the blood was raised to a high level (Fig. 4, line c). This value of V was estimated by *Daniel, Moorhouse,* and *Pratt* (1978 c) from the results of experiments in which the concentration of tryptophan in the blood was raised to successively higher levels. The three lines

Table 1. *Shows how the carrier-mediated transport of tryptophan across the blood-brain barrier is inhibited by other amino acids normally present in the bloodstream. For each amino acid are shown an estimate (± S.E.) of the value of* K_i*, the equilibrium constant for combination of the amino acid with the tryptophan carrier; the mean value of i (± S.E. of mean), i.e. the normal concentration of the amino acid in the blood, as found in a series of 20 rats; the ratio* i/K_i *and values of* K_t*, the Michaelis constant, i.e. the concentration of the amino acid at which it half saturates its own transport across the blood-brain barrier (in the absence of all inhibitors)*

	K_i mmole l^{-1}	i mmole l^{-1}	$\frac{i}{K_i}$	K_t* (intra carotid injection) mmole l^{-1}	K_t+ (calculated from the apparent K_t) mmole l^{-1}
Phenylalanine	0.089 ± 0.016	0.052 ± 0.002	0.59	0.12	0.08 ± 0.02
Leucine	0.14	0.134 ± 0.006	0.96	0.15	0.12 ± 0.002
Tyrosine	0.15 ± 0.03	0.053 ± 0.004	0.35	0.16	0.13 ± 0.03
Histidine	0.13	0.089 ± 0.010	0.68	0.28	0.17 ± 0.02
Valine	0.20 ± 0.03	0.171 ± 0.005	0.86	0.63	1.3 ± 0.3
Isoleucine	0.36 ± 0.06	0.087 ± 0.005	0.24	0.33	0.54 ± 0.09
Arginine	0.58 ± 0.20	0.138 ± 0.020	0.24	0.09	0.84 ± 0.02
Methionine	0.61 ± 0.33	0.041 ± 0.003	0.067	0.19	0.15 ± 0.01
Alanine	0.72 ± 0.20	0.405 ± 0.050	0.56	—	—
Threonine	1.76	0.296 ± 0.021	0.17	0.73	1.2 ± 0.4
Lysine	2.1 ± 0.6	0.363 ± 0.035	0.17	0.10	1.7 ± 0.3
Serine	6.2	0.272 ± 0.016	0.044	—	0.49 ± 0.11
Glycine	8.9 ± 3.4	0.359 ± 0.040	0.040	—	—
Glutamine	9.3	0.967 ± 0.045	0.10	—	—

* *Pardridge* and *Oldendorf* (1975).

\+ Calculated from the data of *Baños, Daniel,* and *Pratt* (1978); *Crockett, Daniel,* and *Pratt* (1978); *Daniel, Moorhouse,* and *Pratt* (1978 c); and *Daniel, Pratt,* and *Wilson* (1977 a, b, c).

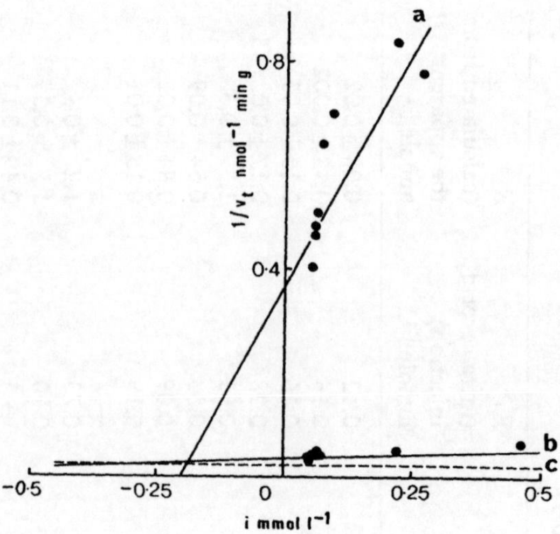

Fig. 3. The effect of raising the concentration, i, of phenylalanine in the circulation upon the carrier-mediated influx of tryptophan into the brain, $\frac{1}{v}$, plotted for two different levels of tryptophan in the blood plasma, *i.e.* 0.07 mmole l^{-1} (line a) and 0.88 mmole l^{-1} (line b). Lines have been fitted to each set of points, using equation (7). Based on data of *Daniel, Moorhouse,* and *Pratt*

intersected in a small area (shown enlarged in Fig. 4). At the point of intersection of any pair of lines $\frac{1}{v_t}$ will be the same and it follows from equation (7) that:

$$\frac{s_1 + K_t\,(1 + \Sigma I_n)}{Vs_1} + \frac{K_t}{K_{i_x} Vs_1}(i' - i_x) = \frac{s_2 + K_t\,(1 + \Sigma I_n)}{Vs_2} + \frac{K_t}{K_{i_x} Vs_2}(i' - i_x) \tag{9}$$

when s_1 and s_2 are the concentrations of tryptophan in each series of experiments and i′ is the concentration of the inhibitor, phenylalanine, at the point of the intersection, which will be the same for each series.

Equation (9) can be re-written as:

$$1 + \frac{K_t}{s_1}\left(1 + \Sigma I_n + \frac{(i' - i_x)}{K_{i_x}}\right) = 1 + \frac{K_t}{s_2}\left(1 + \Sigma I_n + \frac{(i' - i_x)}{K_{i_x}}\right). \tag{10}$$

If $s_1 \neq s_2$ this can only be true if

$$1 + \Sigma I_n + \frac{(i' - i_x)}{K_{i_x}} = 0$$

that is if

$$K_{i_x}(1 + \Sigma I_n) = i_x - i'. \tag{11}$$

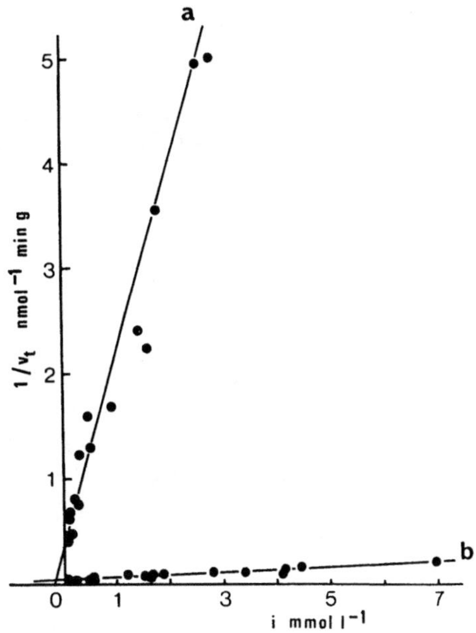

Fig. 4. Shows the portion of Fig. 3 near the origin, enlarged, together with a line (c) representing the reciprocal of the influx, $\frac{1}{V}$, when the concentration of tryptophan is raised to a level high enough to saturate the carrier. Note all three lines meet approximately at a point, the co-ordinates of which are $\frac{1}{V}$ and i′. Such an intersection is in accordance with the kinetics of competitive inhibition [*Dixon*, 1953; see equation (11)]

Thus the negative co-ordinate of the point of intersection of a pair of lines (Fig. 4) provides an estimate of $K_{i_x}(1 + \Sigma I_n)$. Since ΣI_n must be positive, an upper limit was obtained for the value of K_i for phenylalanine, and a reasonable guess was made as to its actual value. This result was used to estimate a value of K_t for tryptophan transport and also to provide a value for i/K_i, with phenylalanine (or other amino acids) as an inhibitor using the data of *Daniel, Moorhouse*, and *Pratt* (1978 c), *Crockett, Daniel*, and *Pratt* (1978), and *Daniel, Love*, and *Pratt* (1978). In order to obtain consistent results it was necessary to assume that s stood for the total concentration of tryptophan in the blood plasma (rather than just for the unbound fraction).

An alternative method was also used, a modification of the procedure of *Hunter* and *Downs* (1945). If v represents carrier-mediated influx, when inhibitor is not present in the blood and v_i represents influx in the presence of a known concentration of the inhibitor, i, then a function

$$i\left(\frac{v_i}{v - v_i}\right)$$

can be calculated. This function is dependant only upon the concentration in the blood of the substance which is transported, s, as given by

$$\frac{v_i}{v - v_i} = K_i + \frac{K_i}{K_t}\, s$$

and estimates of K_i and K_t can be obtained by using this equation.

This *Hunter* and *Downs* (1945) procedure was modified as follows to take account of the various inhibitors of tryptophan transport which are normally present in the bloodstream. It is clear that in life the concentration of the inhibitor will not be reduced below i_x, the level at which it is normally present in the blood. If the influx is then v_x and if the concentration of the inhibitor in the blood is raised to i, the influx is reduced to v_i and from equation (7) the difference in influx will be given by:

$$(v_x - sD) - (v_i - sD) =$$

$$= v_x - v_i = \frac{sV}{K_t(1 + \Sigma I_n) + s} - \frac{sV}{K_t(1 + \Sigma I_n) + s + (i - i_x)\frac{K_t}{K_{i_x}}} \quad (12)$$

provided that the concentration of tryptophan in the blood, s, is unchanged.

This equation can be simplified to

$$v_x - v_i = \frac{sVK_t(i - i_x)}{K_{i_x}(K_t(1 + \Sigma I_n) + s)\left(K_t(1 + \Sigma I_n) + s + (i - i_x)\frac{K_t}{K_{i_x}}\right)}. \quad (13)$$

If equation (13) is multiplied by equation (7) we get

$$\frac{v_x - v_i}{v_i - sD} = \frac{K_t(i - i_x)}{K_{i_x} K_t(1 + \Sigma I_n) + sK_{i_x}} \quad (14)$$

which can be re-written as

$$\frac{v_x - v_i}{(i - i_x)(v_i - sD)} = \frac{K_t}{K_{i_x} K_t(1 + \Sigma I_n) + sK_{i_x}}. \quad (15)$$

If this equation is inverted we get

$$(i - i_x)\frac{v_i - sD}{v_x - v_i} = K_{i_x}(1 + \Sigma I_n) + \frac{K_{i_x}}{K_t} s. \quad (16)$$

The data from each series of experiments on the inhibition of tryptophan influx by raising the level in the blood of another amino acid were analysed using equation (16). From a set of results with different values of i and similar values of s, a mean value of the function

$$(i - i_x)\frac{v_i - sD}{v_x - v_i}$$

was obtained and the procedure was repeated for one or more different values of s. The linear regression of the function upon s was calculated by the method of least squares; the standard errors of the parameters defining the fitted straight line were also evaluated. From equation (16), the slope of the line gave an estimate of the ratio K_i to K_t and the intercept upon the vertical axis gave an estimate of the apparent inhibitor constant K_i which is larger than the real inhibitor constant by a factor $(1 + \Sigma I_n)$. It should be noted that this factor will be the same for each of the different inhibitors.

From the estimates of K_i obtained by these two different methods, based upon the *Dixon* (1953) and the *Hunter* and *Downs* (1945) procedures, a series of estimates of K_i and therefore of $\frac{i}{K_i}$ were obtained for each of the other amino acids present in the bloodstream which are likely to inhibit the transport of tryptophan into the brain. From these results a more accurate estimate was obtained of the value of ΣI_n of 5.05, and these results were used to re-calculate more accurate values of K_i for each of the inhibitors. Thus values for each K_i, for K_t and for ΣI_n which satisfied equations 7, 8 and 16 were obtained. In many cases the value of K_i obtained (Table 1) is of a similar order to that of K_m calculated from our data or reported by *Pardridge* and *Oldendorf* (1975). Tryptophan influx will be especially sensitive to the blood levels of other aromatic acids. By substituting the appropriate values in equation (5) and differentiating, it can be calculated that under normal conditions tryptophan influx into the brain will fall at the rate of about 2 % per 10 μmole l^{-1} increase in the level of phenylalanine, as can be confirmed by examining Fig. 2.

This kinetic analysis of the transport of tryptophan across the blood-brain barrier should help to elucidate the effects of various hormonal changes or metabolic disturbances which may affect the supply of tryptophan to the brain; *e.g.* diabetes, starvation, hepatic failure or inborn errors of amino acid metabolism, all of which conditions alter the proportions of the different amino acids in the circulation. Thus insulin, either given directly or secreted after a carbohydrate meal, has two opposing effects. Firstly, it reduces the levels of free fatty acids in the blood and so causes an increased proportion of the tryptophan in the blood to become bound to the albumin (*Curzon, Friedl,* and *Knott,* 1973). Secondly, it reduces the blood levels of other amino acids which compete with tryptophan for the transport carrier (*Fernstrom* and *Wurtman*, 1972; *Fernstrom, Larin*, and *Wurtman*, 1973). Since there is now general agreement that insulin increases the rate at which tryptophan enters the brain (*Fernando, Knott*, and *Curzon*, 1976; *Tagliamonte et al.*, 1976; *Daniel, Love, Moorhouse*, and *Pratt*, 1979) it seems clear that the *increase* in influx due to the removal of competing amino acids predominates over any *reduction* in influx due to increased binding of tryptophan by blood albumin. Moreover, *Fernstrom, Hirsch*, and *Faller* (1976) have suggested that the transport carrier is able to dissociate the tryptophan-albumin complex while the blood is passing through the cerebral capillaries, which would reduce the importance of this binding. The effect of insulin upon the competing amino acids is important. A rise in tryptophan influx was shown by *Daniel, Love, Moorhouse*, and *Pratt* (1979) to accompany the falls in the levels of blood amino acids caused by raising the level of insulin in the blood (but maintaining normal blood glucose levels) for two hours. The reduction in the levels of inhibitor amino acids which we report should increase (equation 5, Table 1) the influx of tryptophan into the brain some 30 %. It is also clear from Table 1, Fig. 2 and equation (5) that not only is tryptophan influx into the brain considerably increased as a result of the reduction in amino acid levels in the blood caused by insulin, but a quite moderate increase in the circulating levels of one of the other aromatic acids will reduce the carrier mediated transport of tryptophan across the blood-brain barrier to almost negligible proportions. That the brain continues to function at all in such circumstances must be due to the non-saturable component of the transport which ensures that a basic minimum of tryptophan reaches the brain. Under such conditions the supply of tryptophan to the brain will be more directly dependent upon the circulating level of tryptophan itself (equation 7), a conclusion with obvious implications in the treatment of aminoacidaemias (*Pratt*, 1978).

Acknowledgements

I wish to thank my son, F. L. Pratt, for help with mathematical work and the Kennedy Trust for financial assistance.

References

Baños, G., Daniel, P. M., Pratt, O. E.: The effect of age upon the entry of some amino acids into the brain, and their incorporation into cerebral protein. Develop. Med. Child Neurol. *20*, 335—346 (1978).

Crockett, M. E., Daniel, P. M., Pratt, O. E.: Interaction between methionine and the branched-chain amino acids in transport across the blood-brain barrier (submitted 1978).

Curzon, G., Friedl, J., Knott, P. J.: The effects of fatty acids on the binding of tryptophan to plasma proteins. Nature *242*, 198—200 (1973).

Daniel, P. M., Donaldson, J., Pratt, O. E.: The rapid achievement and maintenance of a steady level of an injected substance in the blood plasma. J. Physiol. (Lond). *237*, 8—9p (1974).

Daniel, P. M., Donaldson, J., Pratt, O. E.: A method for injecting substances into the circulation to reach rapidly and to maintain a steady level. Med. Biol. Engineering *13*, 214—227 (1975).

Daniel, P. M., Donaldson, J., Pratt, O. E.: Infusion schedules for prescribed blood concentration time courses. J. appl. Physiol. *41*, 608 (1976).

Daniel, P. M., Love, E. R., Moorhouse, S. R., Pratt, O. E.: Effect of insulin upon levels of amino acids in the blood and upon the influx of tryptophan into the brain. J. Physiol. (Lond.) (in press, 1979).

Daniel, P. M., Love, E. R., Moorhouse, S. R., Pratt, O. E., Wilson, P.: A method for rapidly washing the blood out of an organ or tissue of the anaesthetized living animal. J. Physiol. (Lond.) *237*, 11—12p (1974).

Daniel, P. M., Love, E. R., Pratt, O. E.: The effect of non-aromatic amino acids normally found in the blood upon the transport of tryptophan across the blood-brain barrier. Unpublished observations (1978).

Daniel, P. M., Moorhouse, S. R., Pratt, O. E.: Amino acid precursors of monoamine neurotransmitters and some factors influencing their supply to the brain. Psychological Med. *6*, 277—286 (1976).

Daniel, P. M., Moorhouse, S. R., Pratt, O. E.: Partial exclusion of tryptophan from the brain due to saturation of the transport carrier. J. Physiol. (Lond.) *282*, 9—10p (1978 a).

Daniel, P. M., Moorhouse, S. R., Pratt, O. E.: Post-natal changes in the transport of aromatic amino acids into the brain. J. Physiol. (Lond.) (in press, 1978 b).

Daniel, P. M., Moorhouse, S. R., Pratt, O. E.: The exclusion of L-tryptophan from the brain by raised levels of other aromatic amino acids in the circulation and a way in which this exclusion can be overcome. Submitted (1978 c).

Daniel, P. M., Pratt, O. E., Wilson, P. A.: The transport of L-leucine into the brain of the rat *in vivo:* saturable and non-saturable components of influx. Proc. Roy. Soc. (Lond.) *B 196*, 333—346 (1977 a).

Daniel, P. M., Pratt, O. E., Wilson, P. A.: The exclusion of L-isoleucine or of L-leucine from the brain of the rat, caused by raised levels of L-valine in the circulation, and the manner in which this exclusion can be partially overcome. J. neurol. Sci. *31*, 421—431 (1977 b).

Daniel, P. M., Pratt, O. E., Wilson, P. A.: The influx of isoleucine into the cerebral hemispheres and cerebellum: carrier-mediated transport and diffusion. Quart. J. exp. Physiol. *62*, 163—173 (1977 c).

Dixon, M.: The determination of enzyme inhibitor constants. Biochem. J. *55*, 170—171 (1953).

Dixon, M., Webb, E. C.: Enzymes, 2nd ed., pp. 318—320. London: Longmans. 1964.

Donaldson, J., Pratt, O. E.: A method for displaying the effect of altering the constants of a function and an application to the problem of maintaining steady blood concentrations. J. Physiol. (Lond.) *252*, 5—6p (1975).

Fernando, J. C. R., Knott, P. J., Curzon, G.: The relevance of *both* plasma free tryptophan and insulin to rat brain tryptophan concentration. J. Neurochem. *27*, 343—345 (1976).

Fernstrom, J. D., Hirsch, M. J., Faller, D. V.: Tryptophan concentrations in rat brain. Biochem. J. *160*, 589—595 (1976).

Fernstrom, J. D., Larin, F., Wurtman, R. J.: Correlation between brain tryptophan and plasma neutral amino acid levels following food conversion in rats. Life Sci. *13*, 517—524 (1973).

Fernstrom, J. D., Wurtman, R. J.: Brain serotonin content-physiological regulation by plasma neutral amino acids. Science *178*, 414—416 (1972).

Hunter, A., Downs, C. E.: The inhibition of arginase by amino acids. J. biol. Chem. *157*, 427—446 (1945).

Mood, A. M., Graybill, F. A.: Introduction to the theory of statistics, 2nd ed., pp. 152—164. New York: McGraw-Hill. 1963.

Pardridge, W. M.: Regulation of amino acid availability to the brain. In: Nutrition and the Brain (*Wurtman, R. J., Wurtman, J. J.*, eds.), Vol. 1. New York: Raven Press. 1977.

Pardridge, W. M., Oldendorf, W. H.: Kinetic analysis of blood-brain barrier transport of amino acids. Biochim. Biophys. Acta *401*, 128—136 (1975).

Pratt, O. E.: An electronically controlled syringe drive for giving an injection at a variable rate according to a preset programme. J. Physiol. (Lond.) *237*, 5—6p (1975).

Pratt, O. E.: The transport of metabolizable substances into the living brain. In: Transport Phenomena in the Nervous System: Physiological and Pathological Aspects (Proc. Satellite Symp. Int. Congr. Neurochem., Padua, 1975), pp. 55—75p. New York-London: Plenum Press. 1976.

Pratt, O. E.: Adequate nutrition of the developing brain. In: Advances in Perinatal Neurology, Vol. 1 (*Korobkin, R., Guilleminault, C.*, eds.). New York: Spectrum Publications. 1978, in press.

Tagliamonte, A., de Montis, M. G., Olianas, M., Onali, P. L., Gessa, G. L.: Possible role of insulin in the transport of tyrosine and tryptophan from blood to brain. Adv. Exp. Med. Biol. *69*, 89—94 (1976).

Wilkinson, G. N.: Statistical estimations in enzyme kinetics. Biochem. J. *80*, 324—332 (1961).

Author's address: Dr. O. *E. Pratt*, Department of Neuropathology, Institute of Psychiatry, De Crespigny Park, Denmark Hill, London, SE5 8AF, U.K.

Journal of Neural Transmission, Suppl. 15, 43—54 (1979)

The Role of Blood-Brain Barrier Transport of Tryptophan and Other Neutral Amino Acids in the Regulation of Substrate-Limited Pathways of Brain Amino Acid Metabolism*

W. M. Pardridge

Department of Medicine, Division of Endocrinology and Metabolism, UCLA School of Medicine, Los Angeles, California, U.S.A.

With 1 Figure

Summary

Many pathways of essential neutral amino acid metabolism in the CNS are influenced by precursor availability. Since the delivery of circulating amino acids to brain cells is primarily controlled by the rate of amino acid transport through the blood-brain barrier (BBB), pathways of brain amino acid metabolism are ultimately influenced by the activity (K_m, V_{max}) of the BBB neutral amino acid transport system. The K_m of BBB transport is in the 0.1—0.6 mM range, which approximates the physiologic plasma levels and forms the basis of the unusual sensitivity of the brain to competition effects on neutral amino acid transport. Unlike the brain, the K_m of amino acid transport into other organs is in the 1—10 mM range or greater, which frees these tissues from competition effects in the physiologic range of plasma amino acids. Tryptophan circulates 80—90 % bound to albumin; however, the capacity/affinity ratio of the BBB neutral amino acid transport system exceeds the capacity/affinity ratio of albumin binding of tryptophan, which enables the carrier to strip tryptophan off albumin as it traverses the brain capillary. The activity of the BBB neutral amino acid transport system is probably not modulated by insulin, but is influenced by changes in thyroid hormone status; the transport system is also induced in states of hepatic encephalopathy and this induction process is the primary cause of the increased brain tryptophan and serotonin levels in cirrhosis.

---

* Supported by the National Science Foundation (BNS 78-05500).

Substrate-Limited Pathways of Brain Amino Acid Metabolism

The transport of tryptophan or other metabolic substrates through the brain capillary wall, *i.e.*, the blood-brain barrier (BBB), is an important regulatory unit in the control of brain function if the circulating substrate is precursor to a substrate-limited pathway of cerebral metabolism. There are two types of substrate-limited pathways (Table 1); their classification depends on whether the BBB transport step is *rate-limiting* or *rate-affecting* for the overall pathway of brain metabolism. Pathways limited by the transport step are ketone body utilization (*Ruderman et al.*, 1974) or glycolysis under states of either decreased glucose supply (hypoglycemia) or increased glucose demand (seizures, anoxia, salicylism) (*Pardridge* and *Oldendorf*, 1977); pathways limited by the transport step are characterized by immeasureably low precursor substrate concentrations. Conversely, intracellular levels of the essential amino acids such as tryptophan are always measureable, which indicates their metabolism is never limited by the BBB transport step, but rather is limited by an intracellular enzymatic step. However, if the K_m of the rate limiting enzyme is greater than or equal to the intracellular amino acid level, then precursor supply will influence the rate of the reaction and under these conditions, the BBB transport step is rate-affecting. Several pathways of brain amino acid metabolism fall into the second category of substrate-limited pathways (Table 1), *e.g.*, the synthesis of serotonin, catecholamines, histamine, carnosine, and S-adenosyl methionine is influenced by the supply of tryptophan, tyrosine, histidine, and methionine, respectively (Table 2); similarly, the oxidation of the branched-chain amino acids is a function of branched-chain amino acid supply (*Chaplin et al.*, 1976). Conversely, the K_m of the rate-limiting step of leucine incorporation into proteins in the CNS is about 10 μM leucine (*Ames* and *Parks,* 1976), which is much less than the normal intracellular level of leucine. Therefore, protein synthesis is normally independent of precursor supply; however, under conditions of pathologic hyperaminoacidemias, *e.g.*, phenylketonuria, the transport of essential neutral amino acids through the BBB is impaired to such an extent that intracellular levels of amino acids may fall to the 10 μM range and protein synthesis may become substrate-limited.

Since so many pathways of brain amino acid metabolism are substrate-limited, the transport of amino acids through the BBB is as critical to the regulation of brain metabolism as is the respective intracellular rate-limiting enzymatic step.

Table 1. *Substrate-limited pathways of brain metabolism*

Rate-limiting step	(Substrate) K_m Relationship	Intra-cellular (Substrate)	Pathways
Transport	$S_p^a \leq K_{m_T}^b$	low[e]	Glycolysis (abnormal states) ketone utilization
Enzymatic	$S_{IC}^c \leq K_{m_E}^d$	high	Neurotransmitter synthesis amino acid oxidation protein synthesis (abnormal states)

[a] S_p = plasma substrate concentration.
[b] K_{m_T} = K_m (half-saturation constant) of transport step.
[c] S_{IC} = intracellular substrate concentration.
[d] K_{m_E} = K_m of rate-limiting enzymatic step.
[e] Low intracellular substrate concentration is arbitrarily defined as that concentration which occupies 5 % or less of the intracellular water space.

Table 2. *Substrate-limited pathways of brain amino acid metabolism*

Precursor	Brain level (μmol g^{-1})[a]	Rate-limiting enzyme	Substrate K_m (mM)	Product
Tryptophan	0.05	trp hydroxylase	0.05[b]	Serotonin
Tyrosine	0.09	tyr hydroxylase	0.14[c]	Catecholamines
Methionine	0.02	met adenosyl transferase	0.09[d]	S-adenosyl methionine
Histidine	0.06	his decarboxylase	0.40[e]	Histamine
Histidine	0.06	carnosine synthase	—	Carnosine[f]

[a] *Baños et al.* (1973).
[b] *Friedman et al.* (1972).
[c] *Coyle* (1972).
[d] *Matthyse et al.* (1972).
[e] *Schwartz et al.* (1972).
[f] *Chung-Hwang et al.* (1976).

Blood-Brain Barrier Transport: Role of Competition

The Michaelis-Menten kinetics of neutral amino acid transport through the BBB have been determined with a tissue sampling single injection technique in rats (*Pardridge* and *Oldendorf*, 1975; *Pardridge*,

1977 a). As shown in Table 3, the K_m of BBB amino acid transport approximates the plasma amino acid levels, and this close relationship provides the basis for the marked sensitivity of the brain to amino acid competition. The extent of competition for a single transport site may be quantified in the calculation of the apparent K_m or K_m (app), *i.e.*,

$$K_m\,(\mathrm{app}) = K_m\left(1 + \sum\frac{(\mathrm{AA})}{K_m}\right) \tag{1}$$

Substitution of the individual K_m values and plasma levels (AA) into equation (1) results in the K_m (app) values in Table 3. Substitution of (AA), K_m (app) and V_{max} (maximal transport rate) into the Michaelis-Menten equation,

$$V = \frac{V_{max}\,(\mathrm{AA})}{K_m\,(\mathrm{app}) + (\mathrm{AA})} \tag{2}$$

results in predicted influx (v) rates (Table 3). Influx rates predicted from data obtained with the single injection technique correlate highly, $r = 0.81$ (*Pardridge*, 1977 a), with experimentally observed rates determined with a constant infusion technique (*Baños et al.*, 1973).

By inspection of equation (1), it can be seen that if the transport $K_m >> (\mathrm{AA})$, then $K_m\,(\mathrm{app}) = K_m$, and the effects of competition are negligible. The available data indicates the brain is uniquely sensitive to the effects of competition, because the K_m of cell membrane amino acid transport is an order of magnitude higher in other tissues relative to brain and relative to physiologic plasma levels (*Pardridge*, 1977 b). The known K_m values for the essential neutral amino acids are shown in Table 4. Surprisingly, there are no published estimates of the transport K_m for the essential (not synthetic) amino acids for such important tissues as skeletal muscle, heart, or fat. However, indirect data indicate the K_m in tissues such as skeletal muscle must be high since this tissue is not sensitive to competition effects in the physiologic range (*Guroff* and *Udenfriend*, 1962). Although the brain appears to be uniquely sensitive to competition between *circulating* amino acids, competition effects may also occur in the gut lumen, since luminal amino acid levels may exceed 10 mM after a protein-rich meal (*Adibi* and *Mercer*, 1973).

Rates of net uptake of amino acids by brain are obtained by arteriovenous differences. Such data are not readily available for the rat, so data obtained in the isolated, perfused dog brain are shown in Table 3. The net uptake of most amino acids by brain is not statistically different from zero, which reflects the fact that the

Table 3. *Kinetics of blood-brain barrier transport of essential neutral amino acids*

Amino acid	Plasma level[a] (mM)	K_m[b] (mM)	K_m (app)[c] (mM)	V_{max}[b] (nmol $min^{-1}g^{-1}$)	pre influx[c] (nmol $min^{-1}g^{-1}$)	obs influx[a] (nmol $min^{-1}g^{-1}$)	Net uptake[d] (nmol $min^{-1}g^{-1}$)
Phenylalanine	0.05	0.12	0.45	30	3.0	6.5	g
Leucine	0.10	0.15	0.53	33	5.2	6.2	1.8
Tyrosine	0.09	0.16	0.58	46	6.2	5.3	g
Tryptophan	0.10	0.19	0.71	33	4.1	2.5	g
Methionine	0.04	0.19	0.77	33	1.6	1.6	g
Histidine	0.05	0.28	1.1	38	1.6	2.5	g
Isoleucine	0.07	0.33	1.3	57	2.9	1.9	1.0
DOPA	0.004[e]	0.44	1.9	64	0.1	0.1[f]	g
Valine	0.14	0.63	2.5	49	2.5	1.8	g
Threonine	0.19	0.73	3.0	37	2.2	2.3	g

[a] *Baños et al.* (1973).
[b] *Pardridge* and *Oldendorf* (1975).
[c] *Pardridge* (1977 a).
[d] *Betz* and *Gilboe* (1973).
[e] *Birkmayer et al.* (1973); a representative therapeutic plasma level of DOPA.
[f] *Daniel et al.* (1976).
[g] Not significantly different from zero.

Table 4. *Tissue survey of neutral amino acid transport* K_m *(mM)*

Amino acid	Intestinal epithelia[a]	Renal tubule[b]	Exocrine pancreas[c]	Red blood cell[d]	Liver[e]	BBB
Phenylalanine	1			4	4	0.12
Leucine	2			2	6	0.15
Tryptophan		4				0.19
Methionine	5		3	5		0.19
Histidine	6	5				0.28
Valine	3		7	7		0.63

[a] *Larsen et al.* (1964).
[b] *Lingard et al.* (1973); *Chan* and *Huang* (1971).
[c] *Begin* and *Scholefield* (1965).
[d] *Winter* and *Christensen* (1964).
[e] *Le Cam* and *Freychet* (1977).

primary fate of amino acids in brain is protein synthesis. For every amino acid incorporated into proteins, one is released via proteolysis resulting in no net uptake. Furthermore, the rates of neutral amino acid influx into brain (1—10 nmol/min/g) are set at about the same rate as amino acid incorporation into proteins (1—10 nmol/min/g) (see *Pardridge*, 1977 b). The coupling of amino acid transport rates and protein synthesis rates is also seen in liver (*Pardridge*, 1977 c) and may be a general phenomenon. Although aromatic amino acids are used irreversibly in the synthesis of neurotransmitters, these processes proceed at rates less than 50 pmol/min/g (*Neff et al.*, 1971), which are too low to be detected by arteriovenous differences. The branched-chain amino acids, leucine and isoleucine, appear to be an exception to the rule, as the brain utilizes appreciable amounts of these amino acids (Table 3). In vitro data indicate the rate of oxidation of leucine is several-fold greater than the rate of leucine incorporation into proteins (*Chaplin et al.*, 1976). These in vitro observations illustrate the limitations of studies with slices of brain tissue, since the reverse situation is true in vivo, *i.e.*, the rate of leucine metabolism via protein synthesis is about 5-fold the rate of leucine oxidation (*Pardridge*, 1977 b).

One additional note in regard to brain arteriovenous difference data is the problem of extracerebral contamination (*Hertz* and *Bolwig*, 1976). Extracerebral contamination undoubtedly explains the artifactually high rates of net uptake of amino acids recorded for the human brain (*Felig et al.*, 1973). The net uptake data reported in Table 3 is not influenced by extracerebral contamination since these data were obtained with an isolated perfused brain preparation.

Role of Albumin Binding of Tryptophan

Tryptophan, unlike the other circulating amino acids, is 80—90 % bound by albumin (*McMenamy et al.*, 1961). A central dogma of pharmacology is only the free drug enters the tissues from the circulation. While this rule may apply to drugs that penetrate cell membranes by free diffusion, the free drug rule is not necessarily applicable to polar metabolic substrates such as tryptophan, which traverse cell membranes by mediated mechanisms that involve binding to specific carrier proteins. The discrepancy between predictions made by the free drug model and the experimental observations in regard to BBB transport of tryptophan is illustrated in Fig. 1. If only the free amino acid was available for transport, then the inhibition of BBB transport of tryptophan by albumin should conform to predictions made by the law of mass action (Fig. 1). However, there is a marked disparity between predicted and observed results, since 12 % (1.7 mM) fraction V bovine albumin results in only a 16 % inhibition of BBB

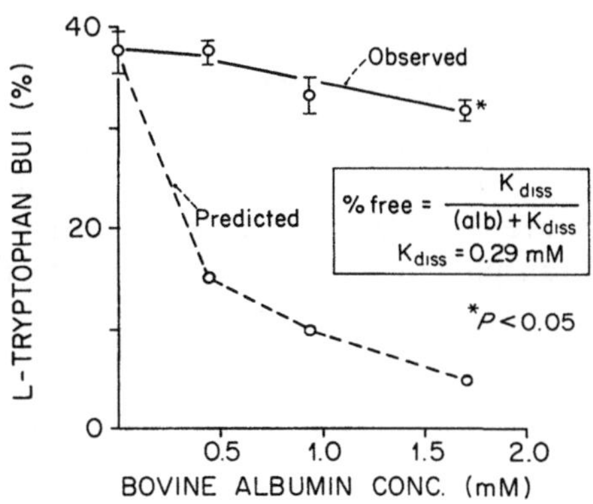

Fig. 1. The brain uptake index (BUI) for a tracer concentration (20 μM) of ^{14}C-L-tryptophan is plotted against the concentration of bovine albumin (fraction V, non-defatted, Sigma Chemical Co.) competing for tryptophan transport into brain. The BUI is a measure of the percent extraction of unidirectional influx of tryptophan through the BBB and is determined with a tissue sampling, single injection technique (*Oldendorf*, 1970). A 200 μl bolus of buffered Ringer's solution containing ^{14}C-L-tryptophan, ^{3}H-water (a highly diffusable internal reference), and various concentrations of bovine albumin is rapidly injected via the common carotid artery in barbiturate-anesthetized rats, followed by decapitation 15 sec later. The % free tryptophan in the respective solutions of albumin may be predicted from the law of mass action as shown in the figure. The dissociation constant for albumin binding of tryptophan ($K_{diss} = 0.29 \pm 0.03$ mM, mean ± SEM, n = 3) was determined by equilibrium dialysis (4 °C, 20 hours). The observed points are mean ± SEM (n = 3—5 rats); $p < 0.05$ refers to statistical difference from the point at zero albumin

transport of tryptophan, as opposed to the 86 % inhibition predicted by the law of mass action (Fig. 1). *Yuwiler et al.* (1977) observed a 22 % inhibition of the BBB transport of tryptophan by 4 % (0.58 mM) defatted bovine albumin; this reflects the greater affinity (lower K_{diss}) of defatted bovine albumin for tryptophan (*Madras et al.*, 1974).

Data such as shown in Fig. 1 and reported elsewhere (*Pardridge*, 1977 a; *Yuwiler et al.*, 1977) suggest the free drug model must be replaced with other models that allow for partition of a ligand (tryptophan) between two competing binding sites (albumin vs the BBB neutral amino acid carrier). An analogous situation is that of a metabolic substrate that is a common precursor to two competing pathways of metabolism. The relative V_{max}/K_m ratio of the two competing enzymes determines the partition of the substrate between the two pathways. Similarly the relative capacity/affinity ratio for albumin and the BBB will determine to what extent the BBB carrier competes for albumin-bound tryptophan. The capacity/affinity ratio for either albumin or the BBB carrier is equal to

$$\frac{\text{(binder concentration)}}{\text{(affinity constant)}}(n) \qquad (3)$$

where n is the number of binding sites per molecule; n = 1 for albumin (*McMenamy et al.*, 1961) and n is assumed to be 1 for the BBB. The data in Fig. 1 indicate the capacity/affinity ratio is much higher for BBB binding and transport of tryptophan relative to albumin binding of the amino acid.

Modulations of the Neutral Amino Acid Carrier

An important area for future investigations is the elaboration of hormonal modulations of the BBB neutral amino acid carrier; such tools would provide a means of altering brain amino acid levels in addition to changes in plasma levels. Insulin is known to modulate the A(alanine)-preferring system in peripheral tissue (*Riggs* and *McKirahan*, 1973), but since the A-system is absent in the BBB (*Wade* and *Katzman*, 1975), it would be expected that BBB transport of amino acids is insulin-independent. Although *De Montis et al.* (1978) suggest insulin increases amino acid transport into brain, an examination of their data fails to confirm this proposal. Insulin was shown to increase the volume of distribution (brain/plasma ratio) of the aromatic amino acids, but this was probably due to the hypo-branched-chain aminoacidemia induced by insulin; since insulin administration failed to also increase the volume of distribution of the branched-chain amino acids (*De Montis et al.*, 1978), it seems unlikely that a generalized induction of the neutral amino acid transport system is occurring as a result of insulin administration.

Thyroid hormone is believed to increase the activity of the L(leucine)-preferring system (*Riggs*, 1970), which is present in the BBB (*Wade* and *Katzman*, 1975). Recent studies indicate that thyroid hormones do modulate BBB amino acid transport (*Daniel et al.*, 1975) and this activity may explain the increased brain serotonin synthesis of hyperthyroidism and the decreased serotonin synthesis of hypothyroidism (*Jacoby et al.*, 1975).

Finally, the activity (V_{max}) of the BBB neutral amino acid carrier is increased in hepatic encephalopathy (*James et al.*, 1978). This important observation would appear to explain the elevated brain tryptophan and serotonin levels of cirrhosis (*Curzon et al.*, 1975). Since the ratio of tryptophan to sum of competing neutral amino acid (a measure of competition) is unchanged in cirrhosis (*Cummings et al.*, 1976), the induction of the BBB amino acid carrier would appear to be the primary mechanism underlying the rise in brain tryptophan and serotonin in hepatic encephalopathy.

References

Adibi, S. A., Mercer, D. W.: Protein digestion in human intestine as reflected in luminal, mucosal and plasma amino acid concentrations after meals. J. Clin. Invest. *52*, 1586—1594 (1973).

Ames, A., III, Parks, J. M.: Functional homogeneity of leucine pool in retina cells. J. Neurochem. *27*, 1017—1025 (1976).

Baños, G., Daniel, P. M., Moorhouse, S. R., Pratt, O. E.: The influx of amino acids into the brain of the rat in vivo: The essential compared with some non-essential amino acids. Proc. R. Soc. Lond. (Biol.) *183*, 59—70 (1973).

Begin, N., Scholefield, P. G.: The uptake of amino acids by mouse pancreas in vitro. II. The specificity of the carrier systems. J. Biol. Chem. *240*, 332—338 (1965).

Betz, A. L., Gilboe, D. D.: Effect of pentobarbital on amino acid and urea flux in the isolated dog brain. Am. J. Physiol. *224*, 580—587 (1973).

Birkmayer, W., Danielczyk, W., Neumayer, E., Riederer, P.: L-Dopa level in plasma, primary condition for the kinetic effect. J. Neural Transm. *34*, 133—143 (1973).

Chan, Y. Z., Huang, K. C.: Microperfusion studies on renal tubular transport of tryptophan derivatives in rats. Am. J. Physiol. *221*, 575—579 (1971).

Chaplin, E. R., Goldberg, A. L., Diamond, I.: Leucine oxidation in brain slices and nerve endings. J. Neurochem. *26*, 701—707 (1976).

Chung-Hwang, E., Khurana, H., Fisher, H.: The effect of dietary histidine level on the carnosine concentration of rat olfactory bulbs. J. Neurochem. *26*, 1087—1091 (1976).

Coyle, J. T.: Tyrosine hydroxylase in rat brain-cofactor requirements, regional and subcellular distribution. Biochem. Pharmacol. *21*, 1935 to 1942 (1972).

Cummings, M. G., Soeters, P. B., James, J. H., Keane, J. M., Fischer, J. E.: Regional brain indoleamine metabolism following chronic portacaval anastomosis in the rat. J. Neurochem. *27*, 501—509 (1976).

Curzon, G., Kantamaneni, B. D., Fernando, J. C., Woods, M. S., Cavanagh, J. B.: Effects of chronic porto-caval anastomosis on brain tryptophan, tyrosine and 5-hydroxytryptamine. J. Neurochem. *24*, 1065—1070 (1975).

Daniel, P. M., Love, E. R., Pratt, O. E.: Hypothyroidism and amino acid entry into brain and muscle. Lancet 2, 872 (1975).

Daniel, P. M., Moorhouse, S. R., Pratt, O. E.: Do changes in blood levels of other aromatic amino acids influence levodopa therapy? The Lancet *1*, 95 (1976).

De Montis, M. G., Olianas, M. C., Haber, B., Taglamonte, A.: Increase in large neutral amino acid transport into brain by insulin. J. Neurochem. *30*, 121—124 (1978).

Felig, P., Wahren, J., Ahlborg, G.: Uptake of individual amino acids by the human brain. Proc. Soc. Exp. Biol. Med. *142*, 230—231 (1973).

Friedman, P. A., Kappelman, A. H., Kaufman, S.: Partial purification and characterization of tryptophan hydroxylase from rabbit hindbrain. J. Biol. Chem. *247*, 4165—4173 (1972).

Guroff, G., Udenfriend, S.: Studies on aromatic amino acid uptake by rat brain in vivo. J. Biol. Chem. *237*, 803—806 (1962).

Hertz, M. M., Bolwig, T. G.: Blood-brain barrier studies in the rat: An indicator dilution technique with tracer sodium as an internal standard for estimation of extracerebral contamination. Brain Res. *107*, 333—343 (1976).

Jacoby, J. H., Mueller, G., Wurtman, R. J.: Thyroid state and brain monoamine metabolism. Endocrinology *97*, 1332—1335 (1975).

James, J. H., Escourrou, J., Fischer, J. E.: Blood brain neutral amino acid transport activity is increased after portacaval anastomosis. Science *200*, 1395—1397 (1978).

Larsen, P. R., Ross, J. E., Tapley, D. F.: Transport of neutral, dibasic and N-methyl substituted amino acids by rat intestine. Biochim. Biophys. Acta *88*, 570—577 (1964).

Le Cam, A., Freychet, P.: Neutral amino acid transport. Characterization of the A and L system in isolated rat hepatocytes. J. Biol. Chem. *252*, 148—156 (1977).

Lingard, J., Rumrich, G., Young, J. A.: Kinetics of L-histidine transport in the proximal convolution of rat nephron studied using the stationary microperfusion technique. Pfluegers Arch. *342*, 13—28 (1973).

Madras, B. K., Cohen, E. L., Messing, R., Munro, H. N., Wurtman, R. J.: Revelance of free tryptophan in serum to tissue tryptophan concentrations. Metabolism *23*, 1107—1116 (1974).

Matthyse, S., Baldessarini, R. J., Vogt, M.: Methionine adenosyltransferase: A double-isotope derivative, enzymatic assay. Anal. Biochem. *48*, 410 to 421 (1972).

McMenamy, R. H., Lund, C. C., Van Marche, J., Oncley, J. L.: The binding of L-tryptophan in human plasma at 37 °C. Arch. Biochem. Biophys. *93*, 135—139 (1961).

Neff, N. H., Spano, P. F., Groppetti, A., Wang, C. T., Costa, E.: A simple procedure for calculating the synthesis rate of norepinephrine, dopamine, and serotonin in rat brain. J. Pharmacol. Exp. Ther. *176*, 701—710 (1971).

Oldendorf, W. H.: Measurement of brain uptake of radiolabeled substances using a tritiated water internal standard. Brain. Res. *24*, 372—376 (1970).

Pardridge, W. M., Oldendorf, W. H.: Kinetic analysis of blood-brain barrier transport of amino acids. Biochim. Biophys. Acta *401*, 128—136 (1975).

Pardridge, W. M., Oldendorf, W. H.: Transport of metabolic substrates through the blood-brain barrier. J. Neurochem. *28*, 5—12 (1977).

Pardridge, W. M.: Kinetics of competitive inhibition of neutral amino acid transport across the blood-brain barrier. J. Neurochem. *28*, 103—108 (1977 a).

Pardridge, W. M.: Regulation of amino acid availability to the brain. In: Nutrition and the Brain, Vol. 1 (*Wurtman, R. J., Wurtman, J. J.*, eds.), pp. 141—204. New York. 1977 b.

Pardridge, W. M.: Unidirectional influx of glutamine and other neutral amino acids in liver of fed and fasted rat in vivo. Am. J. Physiol. *232*, E492—E496 (1977 c).

Riggs, T. R.: Hormones and transport across cell membranes. In: Biochemical Actions of Hormones, Vol. 1 (*Litwack, G.*, ed.), pp. 157—208. New York. 1970.

Riggs, T. R., McKirahan, K. J.: Action of insulin on transport of L-alanine in rat diaphragm in vitro. J. Biol. Chem. *248*, 6450—6455 (1973).

Ruderman, N. B., Ross, P. S., Berger, M., Goodman, M. N.: Regulation of glucose and ketone-body metabolism in brain of anesthetized rats. Biochem. J. *138*, 1—10 (1974).

Schwartz, J. C., Lampart, C., Rose, C.: Histamine formation in rat brain in vivo: Effects of histidine loads. J. Neurochem. *19*, 801—810 (1972).

Wade, L. A., Katzman, R.: Synthetic amino acids in the nature of L-Dopa transport at the blood-brain barrier. J. Neurochem. *25*, 837—842 (1975).

Winter, C. G., Christensen, H. N.: Migration of amino acids across the membrane of the human erythrocyte. J. Biol. Chem. *239*, 872—878 (1964).

Yuwiler, A., Oldendorf, W. H., Geller, E., Braun, L.: Effect of albumin binding and amino acid competition on tryptophan uptake into brain. J. Neurochem. *28*, 1015—1023 (1977).

Author's address: Dr. *W. M. Pardridge*, UCLA School of Medicine, Department of Medicine, Los Angeles, CA 90024, U.S.A.

Journal of Neural Transmission, Suppl. 15, 55—67 (1979)

Diet-Induced Changes in Plasma Amino Acid Pattern: Effects on the Brain Uptake of Large Neutral Amino Acids, and on Brain Serotonin Synthesis

J. D. Fernstrom

Laboratory of Brain and Metabolism, Program in Neural and Endocrine Regulation, Massachusetts Institute of Technology, Cambridge, Massachusetts, U.S.A.

With 6 Figures

Summary

Tryptophan is transported into brain by a competitive carrier system it shares with such other large neutral amino acids as tyrosine, phenylalanine, leucine, isoleucine, and valine. Physiologic variations in the plasma neutral amino acid pattern (either as a change in plasma tryptophan, or in the plasma concentration of one or more of its competitors) directly alter this competitive process, and thereby modify the uptake of tryptophan into brain. Such variations in tryptophan uptake influence brain tryptophan levels, and thus serotonin synthesis. Food intake, by influencing directly the plasma levels of large neutral amino acids, can therefore predictably modify brain tryptophan uptake and serotonin synthesis.

The effect of food intake on the competitive uptake of tryptophan into brain, and on brain tryptophan levels, has recently been shown not to be limited to this amino acid, but also holds for other large neutral amino acids, and for certain large neutral amino acid drugs (*e.g.*, methyldopa). Hence, following a meal, the brain concentration of any large neutral amino acid appears to depend on how the food modifies the plasma level of that amino acid *relative* to the plasma concentrations of its competitors.

The binding of tryptophan to albumin in blood has also been suggested to influence brain tryptophan uptake. However, this notion has not been sustained by the results of nutritional studies, in which meal-induced changes in brain tryptophan levels were readily shown not to be predicted by the alterations in the size of the serum free tryptophan pool.

Taken together, these data affirm the importance of competitive transport in determining brain tryptophan uptake and levels, but question whether serum albumin binding and the size of the free tryptophan pool function physiologically to modulate brain tryptophan concentrations.

Diet has been known to influence brain serotonin (5-HT) levels for almost 20 years: long before kinetic data became available on tryptophan hydroxylase, numerous investigators had shown that chronic increases or decreases in the availability of tryptophan to the body and brain produced elevations or reductions, respectively, in brain 5-HT (*Zbinden et al.*, 1958; *Gal* and *Drewes*, 1962; *Green et al.*, 1962). Moreover, even before good information was obtained on the K_m of tryptophan hydroxylase for its substrate, a group of Scottish investigators demonstrated that the rapid increases in brain tryptophan levels that followed injection of the amino acid elicited sharp increases in the rate of 5-HT synthesis (*Ashcroft et al.*, 1965). When the kinetic data were finally obtained (*Jéquier et al.*, 1969), what was already suspected became fact. That is, the enzyme catalyzing the rate-limiting step in serotonin (5-HT) biosynthesis, tryptophan hydroxylase, is about half-saturated with substrate at normal brain tryptophan concentrations. Hence, variations in brain tryptophan concentration, by modifying the saturation of tryptophan hydroxylase, directly influence the rate of 5-HT synthesis. More recently, we have observed that physiological variations in the availability of tryptophan to the brain, such as occur following meal ingestion, also lead to predictable alterations in brain serotonin synthesis. Foods whose ingestion elevates brain tryptophan also increase 5-HT synthesis (*Fernstrom* and *Wurtman*, 1971), and those whose consumption reduce brain tryptophan predictably lower brain 5-HT (*Fernstrom et al.*, 1975 a). Although the rate of serotonin (5-HT) synthesis may be modulated by a number of control systems, one such modulation thus seems clear: substrate availability to the brain.

The early dietary experiments and more recent tryptophan injection studies did not explore the process by which tryptophan, added to the blood via the diet or the needle, is taken up into brain. The most simple mechanism would be that brain tryptophan uptake and levels simply reflect plasma tryptophan concentrations (*e.g.*, a simple diffusion-mediated uptake). While this notion certainly appears to hold under very *limited* conditions, attempts at more complex and generalizable hypotheses, based on recent data, have been put forth to explain the process by which plasma tryptophan molecules gain access to the brain, and the ability of any treatment ultimately to influence brain tryptophan levels. The two most-often discussed are: (1) the competitive brain transport system for tryptophan and other large neutral amino acids (LNAA), and (2) the binding of tryptophan to albumin in blood. This article will focus on physiological data tending to support the notion that competitive transport is the most important determinant of brain tryptophan uptake, while plasma tryptophan binding to albumin seems to influence brain tryptophan only to a minimal, if not negligible, extent.

Diet, Competitive Amino Acid Transport Into Brain, and 5-HT Synthesis

We showed several years ago that, in addition to tryptophan injection, the ingestion by fasting rats of a single carbohydrate meal elicited rapid increments in tryptophan levels in plasma and brain, and in brain 5-HT levels and synthesis (*Fernstrom* and *Wurtman*, 1971; *Colmenares et al.*, 1975). Initially, we conjectured that these data supported the notion that brain tryptophan and 5-HT levels reflected directly changes in plasma tryptophan concentration. However, in additional experiments, we noted that the consumption of a protein-containing meal (*e.g.*, 18 % casein) which caused large increases in plasma tryptophan, failed to elevate brain tryptophan or 5-HT concentrations (see *Fernstrom* and *Wurtman*, 1972). Clearly, the uptake of tryptophan into brain did *not* simply reflect plasma tryptophan concentrations. Subsequent studies were designed to assess whether the failure of protein consumption to raise brain tryptophan involved the known competition for brain uptake of tryptophan with other large neutral amino acids (see *Pardridge*, 1977). The results showed that, indeed, brain tryptophan failed to rise because the ingestion of protein induced changes in the plasma concentrations of large neutral amino acids (other than tryptophan) that were quite similar to that in plasma tryptophan itself. The single-most illuminating experiment involved feeding rats meals containing, *in lieu* of 18 % casein, either a mixture of synthetic amino acids formulated to mimic the amino acid composition of casein, or amino acid mixtures that varied slightly from this basic mixture. When fasting animals ingested the "complete" diet, which approximated the amino acid composition of the 18 % casein diet, no increments occurred in brain tryptophan or 5-hydroxyindoles. However, when rats consumed a similar diet, differing from the "complete" diet only by the deletion of the large neutral amino acids (except for tryptophan), enormous increments in brain tryptophan and 5-hydroxyindoles resulted (*Fernstrom* and *Wurtman*, 1972). This result supported the notion that, following food consumption, the large neutral amino acid competitors in the protein (or mix) ultimately contributed to preventing an increase in brain tryptophan uptake in the presence of elevated blood tryptophan levels. These data thus indirectly showed that, physiologically, the uptake of amino acids into brain did depend on competition. A confirmation of these ideas consisted of showing in a similar dietary experiment that the deletion from the "complete" amino acid mix diet of those amino acids thought *not* to compete with tryptophan for brain uptake (*e.g.*, either the acidic or basic amino acids) did not modify the response observed following the ingestion of the complete diet. That is, no increments were observed in brain tryptophan or 5-HT.

These and other experiments (*Fernstrom et al.*, 1973, 1975 a, 1976) led us to propose a working hypothesis to explain how variations in the blood concentrations of large neutral amino acids (including tryptophan), induced by food ingestion or other physiologic events, influence tryptophan uptake into brain and 5-HT synthesis (Fig. 1). For example, when the animal consumes a carbohydrate meal, not only does plasma tryptophan increase, but the secretion of insulin elicits a reduction in the plasma concentrations of the other large neutral amino acids. Thus, tryptophan gains a competitive advantage for uptake, and brain tryptophan and 5-hydroxyindole levels rise. When the rat ingests protein in the diet, however, not only does plasma tryptophan increase, but so too do the plasma concentrations of the other large neutral amino acid competitors. For diets containing greater than 18—24 % casein the relative increase in the plasma concentration of tryptophan and its competitors are approximately equal, and thus no relative change occurs in the competition for brain uptake. Hence, brain tryptophan levels and 5-hydroxyindole synthesis do not change. In Fig. 1, we have summarized these diet-induced changes in the plasma neutral amino acid pattern as a plasma ratio of tryptophan to the sum of its large neutral amino acid competitors. Thus, carbohydrate ingestion raises this ratio and brain tryptophan; protein ingestion (18—24 %) usually does not modify the ratio nor brain tryptophan concentration.

This basic hypothesis for describing tryptophan uptake into brain has been found by us (*Fernstrom et al.*, 1975 a, 1976) and others (*e.g.*, *Ashley* and *Anderson*, 1975) to hold in a number of dietary situations. Moreover, the plasma (or serum) ratio model (*i.e.*, the importance of competition in determining brain amino acid uptake) has now been generalized to most of the large neutral amino acids. That is, following meal ingestion, the changes that occur in the brain levels of individual LNAA other than tryptophan correlate directly with the alterations induced by the meal in the appropriate ratio for each

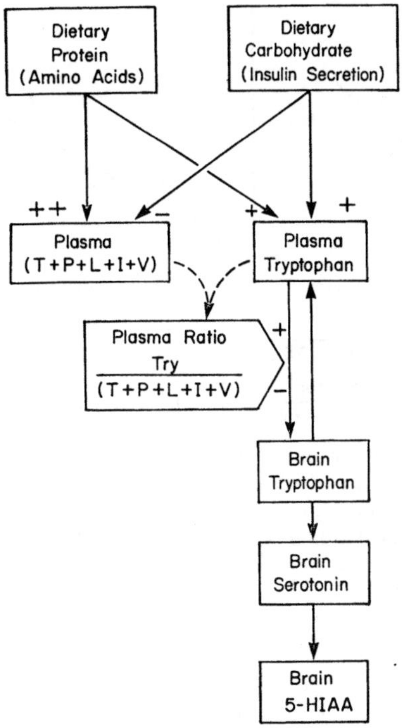

Fig. 1. Proposed sequence describing diet-induced changes in serotonin concentration in the rat. The ratio of tryptophan to the combined levels of tyrosine (T), phenylalanine (P), leucine (L), isoleucine (I), and valine (V) in the plasma is thought to control the tryptophan level in the brain. (*Fernstrom* and *Wurtman*, 1972)

amino acid. For example, in one type of experiment, LNAA levels in blood and brain were measured in animals that consumed for two hours either a 0 %, 18 %, or 40 % casein meal. The effects on brain tryptophan and the serum tryptophan ratio were as anticipated, and the correlation was quite good (left panel, Fig. 2). For valine, another LNAA (Fig. 3), brain levels clearly changed following food ingestion: they fell with the ingestion of the 0 % casein diet, did not deviate significantly from control when 18 % casein was consumed, and rose with the ingestion of the 40 % casein diet. The changes induced by the meals in the plasma valine ratio clearly correlated quite well with the alterations in brain valine levels (Fig. 3, left panel). Similarly good correlations were also found for leucine, tyrosine, and phenylalanine in these experiments (*Fernstrom* and *Faller*, 1978). The ratio was not as predictive of the changes in brain isoleucine, but indeed *was* so when corrected for differences in the affinities of each of the LNAA

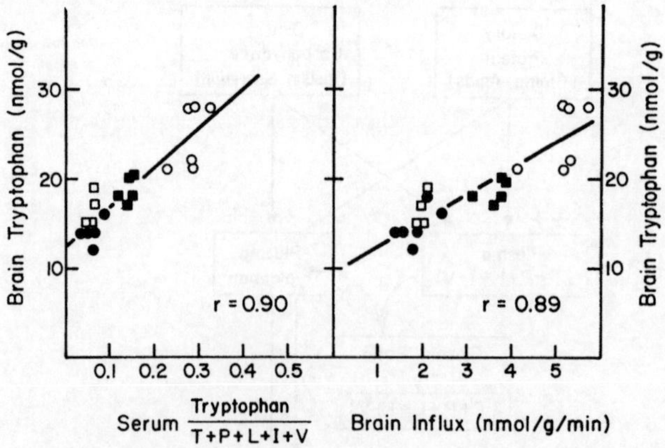

Fig. 2. Relationship between brain tryptophan level and the serum tryptophan ratio, or the brain influx in individual animals consuming single meals of differing protein content for two hours. Closed circles, fasting controls; open circles, rats ingesting 0 % protein; closed squares, rats ingesting 18 % protein; and open squares, rats consuming 40 % protein. The data were analyzed by linear regression; the values of the correlation coefficient (r) are significantly different from 0. $P < 0.01$. (*Fernstrom* and *Faller*, 1978)

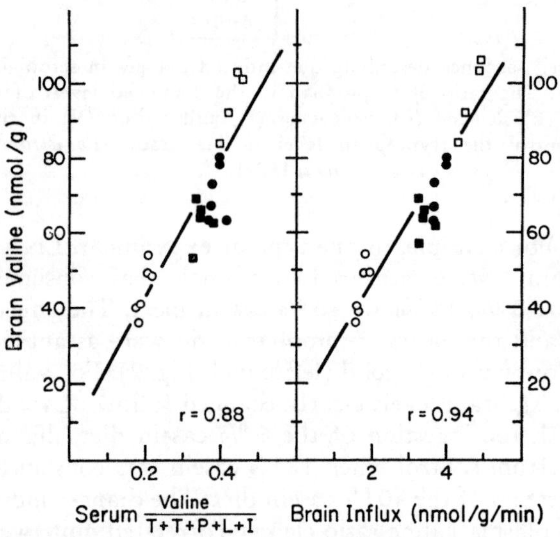

Fig. 3. Relationship between brain valine levels and the serum valine ratio, or brain valine influx in individual rats consuming single meals of differing protein content for two hours. The symbols are the same as those described in Fig. 2; T+T = tryptophan + tyrosine; P = phenylalanine, L = leucine, and I = isoleucine. Both correlation coefficients are significantly different from 0. (*Fernstrom* and *Faller*, 1978)

for the transport carrier (see *Pardridge*, 1977; *Fernstrom* and *Faller*, 1978). For many of the large neutral amino acids, the effect of a particular meal on the brain level of an amino acid (*e.g.*, phenylalanine), which correlated directly with the effect on the plasma ratio (of phenylalanine), did *not* correlate with the alteration in the plasma level of the amino acid by itself. Thus, for example, 0 % casein ingestion elevated brain phenylalanine and the plasma phenylalanine ratio, but lowered serum phenylalanine concentrations (compared to fasting control values). 40 % casein ingestion raised serum phenylalanine, but not the serum ratio, nor brain phenylalanine levels.

The serum (or plasma) ratios do not take into account the fact that each of the LNAA has a unique affinity (K_m) for the competitive transport carrier. *Pardridge* (1977) has described an equation, termed the brain influx, which utilizes the serum concentration of each LNAA and their K_m and V_{max} values (determined by *Pardridge* and *Oldendorf* [1975]), to calculate rates of uptake for each LNAA. We have utilized our data and these K_m and V_{max} values for the rat brain to calculate the brain influx parameter for each LNAA in the study described above. The results for tryptophan and valine are shown in the right-hand panels of Figs. 2 and 3. They indicate an excellent correlation between the calculated brain influx parameters and the brain level of tryptophan and valine following food consumption. The data in Figs. 2 and 3 suggest that the serum ratio is as good a predictor of the brain amino acid level as the calculated influx. Hence, in attempting to predict changes in competitive uptake in species such as the human, in which it is not possible to determine for each amino acid the K_m and V_{max} values for the brain carrier system (the brain must be removed to do so), the data imply that only the serum ratio need be calculated.

Finally, it has recently been possible to obtain yet further evidence indicating the physiologic importance of competitive amino acid transport in determining the brain levels of individual amino acids, and the ultimate functional significance of this phenomenon. We have shown that α-methyl dopa (AMD), a major antihypertensive agent, although a methylated amino acid, nonetheless competes with natural large neutral amino acids for brain uptake in rats (*Markovitz* and *Fernstrom*, 1977). Fig. 4 demonstrates the ability of an injection of LNAA to diminish the uptake of AMD into brain following its administration. This finding led us to pursue dietary studies, and it has been possible to show that the prior ingestion of a high protein meal inhibits the uptake of AMD into brain following its injection, while consumption of a carbohydrate meal enhances brain AMD uptake. This effect of the diet on AMD uptake can be shown to be related to the changes that occur in the plasma levels of the LNAA following food ingestion (carbohydrate lowers them, while 40 % protein consumption elevates them). Using spontaneously hypertensive rats, we then showed that, when animals ingested a meal of carbohydrate prior to AMD injection, AMD caused a large reduction in blood pressure, while in animals consuming a 40 % protein meal, AMD administration produced only a very attenuated reduction in blood pressure (Fig. 5) (*Sved* and *Fernstrom*, 1978). Since aldomet is thought to act within the brain to lower blood pressure, the most economic interpretation of these data is that protein ingestion, by increasing plasma LNAA levels and thereby reducing AMD uptake into brain, also reduced the availability of the drug to the brain sites necessary for its biologic actions. Carbohydrate ingestion, by lowering serum LNAA concentrations and thus enhancing AMD uptake into brain, promoted the fall in pressure, since the drug had ready access to its central site of action. Further studies are now in progress to characterize these effects more fully.

In general, with such continuing lines of investigation it appears that the competitive uptake phenomenon for the transport of LNAA (such as tryptophan) into brain is a genuinely real process, and thus

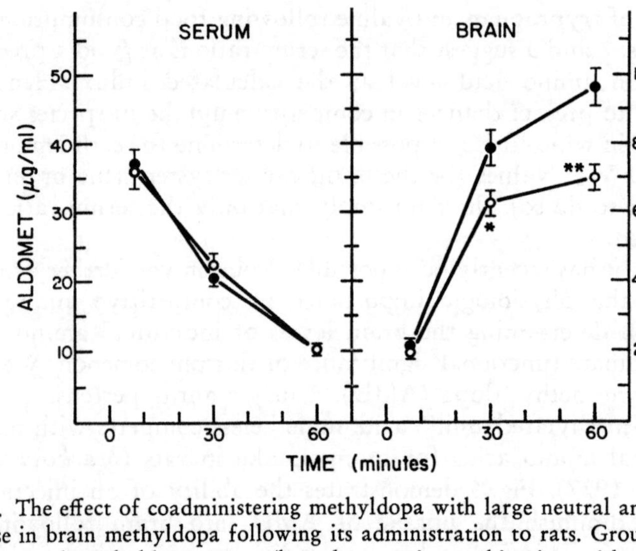

Fig. 4. The effect of coadministering methyldopa with large neutral amino acids on the rise in brain methyldopa following its administration to rats. Groups of fasting rats received methyldopa (50 mg/kg) alone or in combination with a mixture of large neutral amino acids (T, P, L, I, V, 20 mg/kg each, or a total of 100 mg/kg), and were killed at the indicated times thereafter. Vertical bars are standard errors; data were analyzed by analysis of variance, * $P < 0.05$, ** $P < 0.01$. Open circles, methyldopa + neutral amino acids; closed circles, methyldopa alone. (*Markovitz* and *Fernstrom*, 1977)

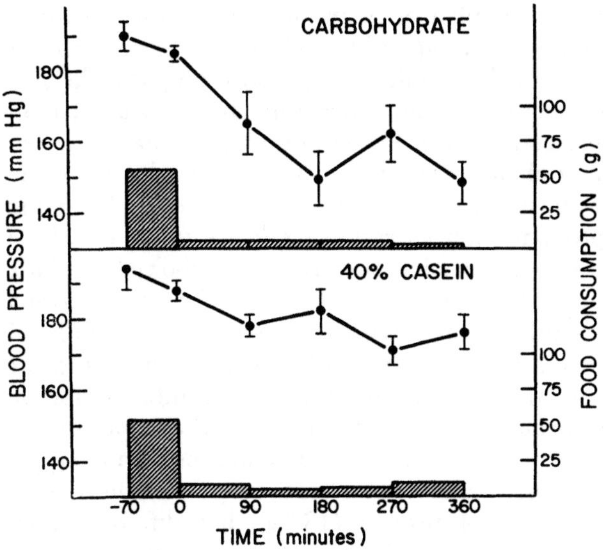

Fig. 5. The effect of food ingestion on the antihypertensive potency of methyldopa in hypertensive rats. Fasted rats (Wistar-Okamoto) consumed either carbohydrate or 40 % casein for 70 min, and then received methyldopa, 50 mg/kg i.p. Blood pressures were measured at the indicated times; food intake was also monitored (hatched bars, representing the intake per group of 6 animals per time interval). Blood pressure was reduced in both groups of animals following methyldopa injection, by the fall was much greater in animals consuming carbohydrate than in those ingesting 40 % casein

that, in predicting the effect of any treatment on brain tryptophan levels (or on the brain uptake and levels of any LNAA), consideration should be given to potential alterations in the plasma concentration of each of the serum LNAA; *i.e.*, the competition among these amino acids for brain uptake.

Diet, Albumin-Binding of Tryptophan in Blood, and Brain Tryptophan Uptake

Tryptophan in serum is distributed between two pools: 10—20 % circulates free, while the remainder binds to serum albumin (*McMenamy* and *Oncley*, 1958). None of the other amino acids binds to blood proteins. Because of this binding phenomenon, several investigators have proposed that the biologically-active moiety of the serum tryptophan pool is that which circulates free, and therefore that the size of this free pool determines the availabality of tryptophan to the brain. This proposition has found some support in the results of a number of pharmacologic studies (*Knott* and *Curzon*, 1972; *Tagliamonte et al.*, 1973). However, in our laboratories, it has been possible to show quite easily that changes in the size of the free tryptophan pool in serum can readily be dissociated from the effects of a given dietary treatment on brain tryptophan. For example, when brain tryptophan levels are increased by the ingestion of carbohydrate, even though serum total tryptophan levels tend to rise, serum *free* tryptophan concentrations *fall* (*Madras et al.*, 1973). Free tryptophan levels decline because the carbohydrate-induced release of insulin lowers the serum concentrations of non-esterified fatty acids (NEFA), which compete with tryptophan for albumin binding sites. Consequently, more tryptophan binds to albumin, and the size of the free tryptophan pool diminishes. Using this information, *Madras et al.* (1974) performed another experiment further to assess the role of the serum free tryptophan pool in modulating brain tryptophan concentrations. They fed groups of rats a series of diets, each containing the same amount of protein (18 %), but different amounts of fat (0, 15, 30, or 45 %) (Fig. 6). The carbohydrate-induced fall in serum NEFA levels can be offset in part, or totally, by the inclusion of fat in the diet. By adding increasing amounts of fat to the diets the rats consumed, *Madras et al.* (1974) generated a dose-response curve relating dietary fat content to serum NEFA levels (Panel D, Fig. 6). Because of this step-wise change in serum NEFA, a similar step-wise increase was also obtained in the size of the serum free (but not total, since *all* groups ate 18 % casein) tryptophan pool (Panel B, Fig. 6). Despite an almost 3-fold range in serum free tryptophan concentrations in these animals (killed 2 hours after food presentation), brain tryptophan levels were almost identical (Panel A, Fig. 6). Clearly, the importance of the free tryptophan pool in determining brain tryptophan levels was negligible in this situation.

Using the Oldendorf technique, *Yuwiler et al.* (1977) have studied simultaneously the importance of albumin-binding and of amino acid competition in determining brain tryptophan uptake. They observed that competition among LNAA had by far the greater impact on tryptophan transport, but that albumin-binding could influence transport to a significant, although small, extent. Studies such as those outlined in Fig. 6 do not support a relationship between serum tryptophan binding and brain tryptophan uptake under normal conditions. Moreover, in other experiments, it has even been possible to show that a serum *free* tryptophan *ratio* is not as reliable as the serum *total* tryptophan ratio in predicting the changes in brain tryptophan that follow food ingestion (*Fernstrom et al.*, 1976). Presumably, the *total* tryptophan ratio is more accurate because of

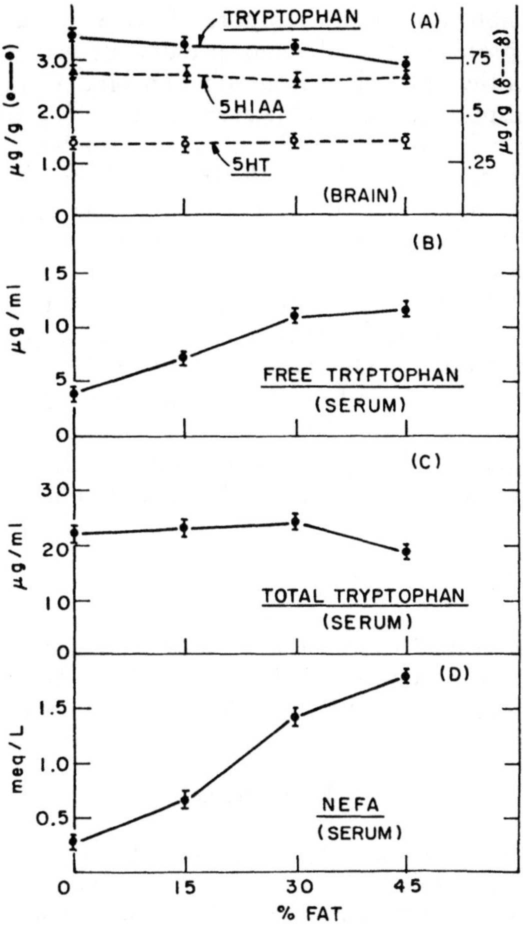

Fig. 6. Effect of dietary fat content on serum tryptophan binding, and on brain indole levels. Groups of 12 fasting rats consumed for two hours one of four diets: all contained 20 % protein (casein), but each differed in fat content. The diets contained 0 % fat, 15 % fat, 30 % fat, or 45 % fat (% weight); carbohydrates supplemented fat on a gram for gram basis as fat was removed from the diet (the diets were thus not isocaloric). A large variation was noted in the serum free tryptophan level among the various groups (panel B), but no significant differences were obtained for *any* of the brain indoles (panel A). (*Madras et al.*, 1974)

the higher affinity of the amino acid for the blood-brain barrier transport carriers than for albumin (*Yuwiler et al.*, 1977). That is, because of the difference in affinities, bound tryptophan molecules can readily be stripped from albumin binding sites as they pass through the brain capillary circulation. Hence, although some tryptophan molecules in blood are bound and some free, almost *all* are readily accessible for brain transport (*Yuwiler et al.*, 1977). (Of course, *all* tryptophan molecules must presumably *become* free, at least transiently, as they dissociate from albumin binding sites and attach to brain transport carriers. However, such a *flux* of tryptophan molecules through the free pool in no manner implies that changes in the *size* of the "transit" pool [*i.e.*, the free pool] should be of value in predicting alterations in brain tryptophan uptake and levels.) Only under certain conditions, then, in which, for example, the administration of pharmacologic agents (*Knott* and *Curzon*, 1972; *Tagliamonte et al.*, 1973) induces a substantial dissociation of tryptophan from albumin, do increments in both serum free tryptophan and brain tryptophan levels occur. However, these latter observations do *not* a generalization make. Therefore, one cannot conclude at the present time, based on available data, that the size of the serum free tryptophan pool (*i.e.*, the binding of tryptophan to albumin) *normally* has any significant impact on the uptake of tryptophan into the brain.

Acknowledgements

The studies summarized in this article were supported in part by grants from the National Institute of Mental Health, NASA, the National Science Foundation, The John A. Hartford Foundation, and the American Heart Association.

References

Ashcroft, G. W., Eccleston, D., Crawford, T. T. B.: 5-Hydroxyindole metabolism in the rat brain: a study of intermediary metabolism using the technique of tryptophan loading. J. Neurochem. *12,* 483—492 (1965).

Colmenares, J. L., Wurtman, R. J., Fernstrom, J. D.: Effect of ingesting a carbohydrate-fat meal on the levels and synthesis of 5-hydroxyindoles in various regions of the rat central nervous system. J. Neurochem. *25,* 825—829 (1975).

Fernstrom, J. D., Faller, D. V.: Neutral amino acids in the brain: changes in response to food ingestion. J. Neurochem. *30,* 1531—1538 (1978).

Fernstrom, J. D., Faller, D. V., Shabshelowitz, H.: Acute reduction of brain serotonin and 5-HIAA following food consumption: correlation with the ratio of serum tryptophan to the sum of competing neutral amino acids. J. Neural Transm. *36,* 113—121 (1975 a).

Fernstrom, J. D., Hirsch, M. J., Faller, D. V.: Tryptophan concentrations in rat brain: failure to correlate with serum free tryptophan, or its ratio to the sum of other serum neutral amino acids. Biochem. J. *160,* 589—595 (1976).

Fernstrom, J. D., Hirsch, M. J., Madras, B. K., Sudarsky, L.: Effects of skim milk, whole milk, and light cream on serum tryptophan binding and brain tryptophan concentrations. J. Nutrition *105*, 1359—1362 (1975 b).

Fernstrom, J. D., Larin, F., Wurtman, R. J.: Correlations between brain tryptophan and plasma neutral amino acid levels following food consumption in rats. Life Sciences *13*, 517—524 (1973).

Fernstrom, J. D., Wurtman, R. J.: Brain serotonin content: increase following ingestion of carbohydrate diet. Science *174*, 1023—1025 (1971).

Fernstrom, J. D., Wurtman, R. J.: Brain serotonin content: physiological regulation by plasma neutral amino acids. Science *178*, 414—416 (1972).

Gal, E. M., Drewes, P. A.: Studies on the metabolism of 5-hydroxytryptamine. Effect of tryptophan deficiency in rats. Proc. Soc. Exptl. Biol. Med. *110*, 368—371 (1962).

Green, H., Greenberg, S. M., Erickson, R. W., Sawyer, J. L., Ellison, T.: Effect of dietary phenylalanine and tryptophan upon rat brain amine levels. J. Pharmacol. Exptl. Therap. *136*, 174—178 (1962).

Jequier, E., Robinson, D. S., Lovenberg, W., Sjoerdsma, A.: Further studies on tryptophan hydroxylase in rat brainstem and beef pineal. Biochem. Pharmacol. *18*, 1071—1081 (1969).

Knott, P. J., Curzon, G.: Free tryptophan in plasma and brain tryptophan metabolism. Nature *239*, 452—453 (1972).

Madras, B. K., Cohen, E. L., Fernstrom, J. D., Larin, F., Munro, H. N., Wurtman, R. J.: Dietary carbohydrate increases brain tryptophan and decreases serum free tryptophan. Nature *244*, 34—35 (1973).

Madras, B. K., Cohen, E. L., Messing, R. B., Munro, H. N., Wurtman, R. J.: Relevance of free tryptophan in serum to tissue tryptophan concentrations. Metabolism *23*, 1107—1116 (1974).

Markovitz, D. C., Fernstrom, J. D.: Diet and uptake of aldomet by brain: competition with natural large neutral amino acids. Science *197*, 1014 to 1015 (1977).

McMenamy, R. H., Oncley, J. L.: The specific binding of L-tryptophan to serum albumin. J. Biol. Chem. *223*, 1436—1440 (1958).

Pardridge, W. M.: Regulation of amino acid availability to the brain. In: Nutrition and the Brain, Vol. I (*Wurtman, R. J., Wurtman, J. J.*, eds.), pp. 141—204. New York: Raven Press. 1977.

Pardridge, W. M., Oldendorf, W. H.: Kinetic analysis of blood-brain barrier transport of amino acids. Biochim. Biophys. Acta *401*, 128—136 (1975).

Sved, A. F., Fernstrom, J. D.: Diet alters the antihypertensive potency of methyldopa. The Pharmacologist *20*, 189 (1978).

Tagliamonte, A., Biggio, G., Vargiu, L., Gessa, G. L.: Free tryptophan in serum controls brain tryptophan level and serotonin synthesis. Life Sci. *12*, 277—287 (1973).

Zbinden, G., Pletscher, A., Studer, A.: Alimentare Beeinflussung der enterochromaffinen Zellen und des 5-Hydroxytryptamingehaltes von Gehirn und Darm. Z. Ges. Exptl. Med. *129*, 615—620 (1958).

Author's address: Prof. *J. D. Fernstrom*, Building 56, Room 137, Massachusetts Institute of Technology, Cambridge, MA 02139, U.S.A.

Journal of Neural Transmission, Suppl. 15, 69—79 (1979)

When—and Why—Should Nutritional State Control Neurotransmitter Synthesis?

R. J. Wurtman

Laboratory of Neuroendocrine Regulation, Department of Nutrition and Food Science, Cambridge, Massachusetts, U.S.A.

Summary

The rates at which neurons synthesize such Group I neurotransmitters as serotonin, acetylcholine, and the catecholamines norepinephrine and dopamine depend physiologically on the availability to them of the circulating precursors for these compounds (tryptophan, choline and tyrosine, respectively). The concentrations of precursor in the circulation and in neurons change rapidly after food consumption, depending upon what is eaten. Nutrient intake thus normally influences the synthesis of these neurotransmitters. Neurons that emit signals by releasing serotonin, acetylcholine, dopamine, or norepinephrine participate in the control of a number of bodily functions and behaviors (*e.g.*, hunger, food choice, sleep, alertness, sensitivity to environmental stimuli and disease states). Dietary manipulations (or the consumption of individual nutrients) can thus be used as tools for the experimental analysis of functions mediated by monoaminergic or cholinergic neurons, and as adjuncts in the treatment of some diseases of these neurons. It is unclear "why" the evolutionary process should have "allowed" the neurotransmission mediated by acetylcholine or the monoamine transmitters to be influenced by the vagaries of food choice. One possible benefit that might accrue to the organism as a result of this dependency would be the use of cholinergic or monoaminergic neurons as "sensors", providing the brain with information about peripheral metabolic state. Thus carbohydrate consumption, which—by altering plasma amino acid levels accelerates brain serotonin synthesis—enhances the release of a transmitter (serotonin) that tends to diminish the animal's desire to consume carbohydrates.

The brain has long been thought to occupy so exalted a position among the organs of the body that its metabolic needs are invariably the first to be met. Thus, if the quantities of oxygen, glucose, or heat present in the blood were in short supply, the scarce item would preferentially be utilized by the brain. An unstated corollary of this formulation is that, under normal conditions, the amounts of circulating nutrients available to the brain at any moment are largely independent of their concentrations in the blood: the brain knows what substances it requires, and takes them from the circulation, oblivious of the dietary history or metabolic state of its host.

My associates at MIT and I have been surprised to discover that this view of the brain as an autonomous organ, largely independent of metabolic processes, is no longer tenable. Our first inkling of the extent to which brain composition can passively reflect peripheral metabolic processes grew out of studies on tryptophan and serotonin, *i.e.*, our assumption that *physiological* changes in plasma amino acid concentration controls brain tryptophan levels, and thereby determines the amount of serotonin synthesized in neurons.

Most recently, we have found that each meal can, depending on its composition, also alter brain choline and tyrosine levels, and thereby modify the synthesis and release of acetylcholine and the catecholamines dopamine or norepinephrine from neurons synthesizing these transmitters. As described below, brain tryptophan and tyrosine levels depend primarily on the proportions of carbohydrate and protein in the meal, while choline levels depend on the diet's lecithin content. We were very surprised when we discovered that serotonin release in the brain normally is affected by food choice; we have been all the more astonished to find similar vulnerabilities among cholinergic and catecholaminergic neurons. We are still a long way from understanding why the evolutionary process apparently allowed the synthesis of a number of important neurotransmitters not to be controlled by the brain. However, we do begin to understand the implications of the nutritional control of at least one transmitter, serotonin, and a number of laboratories, including our own are starting to utilize the brain's dependence on precursor availability to learn more about normal brain function, and to begin to treat disease states related to neurotransmitters whose syntheses are precursor-dependent.

Neurotransmitters

All of the known neurotransmitters synthesized in mammalian neurons share certain chemical properties: all are low molecular weight, water-soluble compounds which are ionized at the pH of body fluids. All are synthesized primarily in portions of the neuron (the synaptic bulbs, or nerve terminals) which form synaptic contacts with the cells to which the neuron transmits signals (the postsynaptic cells), and all are stored within subcellular organelles (synaptic vesicles) or are attached to cytoplasmic proteins, prior to being released into the synaptic cleft. Release occurs when the neuron that synthesized the neurotransmitter is depolarized. Thereafter, some of the liberated molecules come into contact with receptors on the surface of the postsynaptic cells. This interaction ultimately leads to a change in the rate at which a particular ion (*e.g.* sodium or chloride) enters the postsynaptic cell; such activations in ionic fluxes are integrated continuously, and determine whether or not depolarization of the postsynaptic cell will occur. Some of the transmitters (like acetylcholine) are excitatory, in that the changes in ionic fluxes that they produce enhance the likelihood that the postsynaptic cell will become depolarized. Others (like, probably, serotonin) are inhibitory, hyperpolarizing the postsynaptic cell's membrane, and thus diminishing the likelihood that depolarization will occur.

At the present time, about 15—20 compounds have been identified in mammalian central nervous systems that *probably* function as neurotransmitters. These putative transmitters can be classified as falling within one of three groups, depending on their probable modes of synthesis. As we shall see, only for one of these three groups does it seem likely that the rate of neurotransmitter synthesis might normally be coupled to plasma composition and peripheral metabolic phenomena.

Proposed Classification of Neurotransmitters by Control of Synthesis

Group I transmitters include, in addition to serotonin and acetylcholine, the catecholamine transmitters dopamine and norepinephrine, and probably histamine. Each of these compounds is synthesized from a precursor that neurons cannot make, and which thus must be obtained from the circulation. The synthesis of each is catalyzed by an enzyme (*e.g.* tryptophan hydroxylase for serotonin; choline acetyltransferase for acetylcholine; tyrosine hydroxylase for dopamine and norepinephrine) that has a relatively *low* affinity for its substrate, and which thus normally operates at less-than-maximal efficiency, because it virtually never is fully saturated with its particular substrate. It should be noted that tyrosine hydroxylase apparently exists in two states, *i.e.*, an „inactive" form and a possibly-phosphorylated "active" form. The hydroxylation of tyrosine by the "inactive" form is probably limited by the availability of the tetrahydrobiopterin cofactor; in contrast, hydroxylation by the "active" form is more affected by intraneuronal tyrosine levels, and proceeds at a faster rate. The conversion of the "inactive" form of the enzyme to the "active" form apparently occurs when the firing rate of the neuron containing the enzyme has been increased. In contrast to tyrosine hydroxylase, the enzymes that catalyze serotonin and acetylcholine synthesis—tryptophan hydroxylase and choline acetyltransferase—may exist in only a single form whose catalytic activity is always limited by precursor availability.

Group II transmitters include the various peptides that have been found in brain neurons and that are thought by many scientists to function as neurotransmitters, largely because of their presence in neurons and their ability to modify ionic fluxes when applied to neurons. The initial steps in the synthesis of these compounds almost certainly are mediated not by enzymes but by polyribosomes (*i.e.* strands of messenger RNA attached to ribosomes). The large peptides formed on the polyribosomes are probably then subjected to major post-transitional modification by enzymes. The concentrations of amino acids needed to allow polysome-directed peptide synthesis to occur at maximal rates in brain apparently are quite low; hence, it doesn't seem likely that normal, food-induced variations in plasma amino acid levels will be found to have much effect on the formation of the peptide neurotransmitters. However, this remains to be examined.

Group III transmitters probably include some amino acids that can be synthesized by all cells (glycine; glutamate; aspartate) and GABA (gamma-aminobutyric acid), an amino acid formed characteristically in brain neurons from glutamate. Since all neurons are able to synthesize these non-essential amino acids from glucose or from any other energy source, it also seems unlikely that their production as neurotransmitters in nerve terminals will be found to vary normally with food consumption or plasma composition. However, no data are available from experiments examining the nutritional control of their synthesis in the brain, and the possibility that they are under such control must still be kept open. Other putative neurotransmitters that may be included in this category are taurine and beta-alanine.

In order to demonstrate that a particular neurotransmitter falls within Group I, and is thus subject to nutritional control, four types of evidence should be obtained: (A) Brain neurons must not be able to synthesize necessary quantities of the transmitter's precursor from glucose. (B) The brain must be able to obtain the precursor by taking it up from the circulation. The uptake system, localized within the brain capillaries at the site of the blood-brain barrier, should also exhibit only a *low* affinity for the precursor, thereby allowing the rate at which uptake occurs to increase and decrease when plasma levels of the precursor rise and fall. (C) Plasma levels of the precursor must rise and fall (*e.g.* in response to eating, or to fasting, or to the secretion of one or more hormones). (D) The key regulatory enzyme that the neurons use to convert the precursor to the neurotransmitter should also, as described above, have a low affinity for the precursor (and thus normally function at less-than-maximal efficiency). Moreover, no closed feedback loops should operate within the neuron to keep its neurotransmitter levels constant. (For example, if the concentrations of the neurotransmitter within the neuron fed back onto the control enzyme involved in its synthesis, decreasing the enzyme's activity as soon as transmitter concentrations become elevated, then an increase in precursor levels couldn't cause a prolonged acceleration in the synthesis of the neurotransmitter.) These criteria, and the proposed classification of neurotransmitters based on their mode of synthesis, were arrived at only slowly, as it became established that the synthesis of serotonin, and then of acetylcholine and the catecholamines could normally be influenced by precursor availability and nutrient intake.

Serotonin

Our initial studies, started with *Fernstrom* in 1970, were designed to determine whether the increase in brain serotonin levels previously observed in rats given a large dose of tryptophan might be indicative of changes that *normally* occur when plasma tryptophan levels rise, such as after eating (*Fernstrom* and *Wurtman,* 1971 a, 1974). To this end, we first obtained blood and brain samples from animals sacrificed at various times of day and night, to characterize the normal variations in tryptophan and serotonin. We found that, immediately after the onset of darkness, when rats normally begin to eat, parallel elevations occurred in blood and brain tryptophan levels, and, soon thereafter, in brain serotonin concentrations. These findings were interpreted as suggesting that serotonin synthesis in brain neurons might normally be coupled to plasma tryptophan concentration, which in turn changed following the ingestion of food. Further evidence supporting this hypothetical relationship was obtained by injecting rats with very small doses of tryptophan, at a time of day (noon) when blood and brain tryptophan levels were normally lowest, and showing that the increases in brain tryptophan caused by this treatment also were associated with increases in brain serotonin. In subsequent studies, the injection of insulin, which raises plasma tryptophan levels in the rat, was also observd to cause the now-anticipated rise in brain tryptophan and serotonin (*Fernstrom* and *Wurtman*, 1972 a). At this point, we decided to see whether the secretion of the animal's own insulin similarly affected brain serotonin synthesis. We fasted rats overnight, then gave them free access for two hours to a carbohydrate meal which, we knew, would elicit insulin secretion. To our surprise and delight, consumption of this single meal caused sequential elevations in blood tryptophan, brain tryptophan, and brain serotonin levels (*Fernstrom* and *Wurtman*, 1971 b). Thus was it shown that eating, *per se*, could normally affect the synthesis of a brain neurotransmitter.

If a large amount of protein was added to the meal, the resulting changes in both brain tryptophan and brain serotonin levels were the opposite of what we had originally anticipated: both fell. Protein contains tryptophan, and contributes some of its tryptophan molecules to the blood after its ingestion. Thus, we had assumed that the addition of protein to the meal would be associated with an increase in brain tryptophan. The fact that the opposite change occurs was ultimately shown to result from one of the properties of the system that transports circulating tryptophan across the blood-brain barrier and into the brain: This same system also transports numerous other neutral amino acids found in protein (*e.g.* tyrosine, methionine, leucine) (*Pardridge*, 1977). When protein is consumed, the resulting increases in the plasma concentrations of these other amino acids are greater, proportionately, than that of tryptophan. Hence, the more protein in the meal, the lower the ratio in blood of the tryptophan concentration to that of its neutral amino acid competitors, and thus the slower the uptake of tryptophan into the brain (*Fernstrom* and *Wurtman*, 1972 b; *Wurtman* and *Fernstrom*, 1976). Serotonin-containing neurons thus appear to be variable-ratio sensors, modifying the production, and probably the release, of their neurotransmitter as an inverse function of each meal's protein content. These neurons probably function to provide the brain with information about peripheral metabolic state, which the brain can use to decide what and when to eat, whether to sleep, when to cause the secretion of particular hormones, and so forth. Changes in most of these and other brain functions have now been observed following the administration of tryptophan or the consumption of meals that modify brain serotonin (e.g. *Wurtman* and *Wurtman*, 1977). The effects of the diet on plasma amino acid levels (and thus, in all likelihood, serotonin synthesis), in humans appears to be virtually identical with those observed in the rat (*Fernstrom* and *Wurtman*, 1978, in preparation).

Acetylcholine

About four years ago, *Cohen* and I began to explore the relationship between blood choline levels and the synthesis of the neurotransmitter acetylcholine. The enzyme that converts choline to acetylcholine, choline acetyltransferase, was already known to share with tryptophan hydroxylase the property of having a low affinity for its substrate; and thus it probably also was not working at full efficiency at the choline concentrations normally present in the tissues. Hence, it seemed possible that the synthesis of acetylcholine, like that of serotonin, might be accelerated by treatments that increased the levels of its precursor in the brain. To our knowledge, no one had ever previously examined the effects on brain acetylcholine of administering choline to animals. This lacuna probably reflected the great difficulty scientists previously had in examining the effects of *any* treatment on brain acetylcholine levels: within a few seconds of death, a major fraction of the acetylcholine in the rat brain is destroyed by a ubiquitous enzyme, acetylcholinesterase. Hence, even a 25 % increase of decrease in brain acetylcholine concentrations (a rather sizeable change for a neurotransmitter pool) might not have been observable when the brain was examined after death. Our first task was to develop a method for circumventing the post-mortem hydrolysis of brain acetylcholine; this was accomplished by killing the animal with a focussed beam of microwave radiation. This treatment simultaneously coagulates all brain proteins, thereby inactivating acetylcholinesterase (and other brain enzymes).

In short order, we found that the administration of choline, by injection or as a constituent of the diet, caused major sequential elevations in serum choline, brain choline, and brain acetylcholine levels (*Cohen* and *Wurtman,* 1976). Concentrations of the neurotransmitter could be shown to rise within all cholinergic neurons, inside and outside the brain, and in the portion of the neuron (the nerve terminals) from which it is released into synapses. Subsequent studies have shown that blood choline levels in humans and experimental animals *normally* vary as a function of dietary choline content (*Hirsch et al.,* 1978); that the effects of dietary choline provided as lecithin (the form in which it is usually ingested) are, if anything, far greater than those observed when choline itself is administered (*Wurtman et al.,* 1977); that the acetylcholine-forming enzyme is not subject to end-product feedback control; and that increases in acetylcholine *levels* are, in fact, associated with parallel changes in the amounts of the transmitter that are released into synapses to act on postsynaptic receptors (*Ulus et al.,* 1978). Exploration of this latter relationship has been facilitated by the accessibility of the cholinergic nerves located outside of the brain (*e.g.*, those that innervate the adrenal medulla and cause it to secrete epinephrine).

Within four months of the publication of our first article showing that choline administration elevates brain acetylcholine level in rats, the first note appeared in a medical journal suggesting a clinical use for this effect. A letter to the editor of the *New England Journal of Medicine* by a group of psychiatrists working at Stanford University (*Davis et al.*, 1975) described marked improvements after choline administration in a patient suffering from a neurological disease, tardive dyskinesia. This disease causes uncontrolled movements of the tongue, mouth, face, and upper trunk; it is a common side effect associated with the use of all presently-marketed antipsychotic drugs. In some patients, it may continue indefinitely, long after treatment with the antipsychotic drug has been discontinued. Similar movement disorders may also occur spontaneously in patients with no history of having taken antipsychotic drugs. On the basis of largely-indirect evidence, neurologists and psychiatrists had speculated for several years that this movement disorder resulted from the release of inadequate amounts of acetylcholine within the brain. However, no orally-effective compound was available that could be administered chronically to restore normal cholinergic tone within the brain.

Subsequently, choline has been affirmed as a useful therapy for tardive dyskinesia in a controlled, double-blind, cross-over study (*Growdon et al.*, 1977), and oral lecithin has also been shown to exert positive therapeutic effects (*Growdon et al.*, 1978). No other potentially-useful treatment, to our knowledge, has ever been shown in a controlled study to benefit patients with this disease. Even though choline or lecithin administration apparently suppresses a major side-effect of antipsychotic drugs (*e.g.*, tardive dyskinesis), we find no evidence that its administration in any way interferes with the therapeutic actions of these drugs. This dissociation may have important implications in helping us to identify the brain loci involved in causing psychoses.

Similar studies are now underway at MIT and numerous other institutions on the possible uses of choline or lecithin to treat other diseases that are thought to involve cholinergic neurons. These include both brain diseases, such as mania and memory loss, and peripheral disorders, like the myasthenic syndromes. The ability of choline to enhance cholinergic transmission has also proven very useful in basic science studies designed to explore the functions of cholinergic neurons in the brain. In contrast, virtually nothing is known about "why" cholinergic neurons are "open-loop" with respect to precursor, or about the benefits—if any—that normally accrue to the animal as a result of this relationship.

Catecholamines

The production and release of the catecholamine neurotransmitters dopamine and norepinephrine can also be influenced by physiologic variations in brain tyrosine levels. The rate at which brain neurons accumulate DOPA after pharmacological inhibition of DOPA decarboxylase—an index of the rate at which they would otherwise synthesize dopamine or norepinephrine—has been shown to increase when brain tyrosine levels are elevated (by giving rats tyrosine) and to decrease when tyrosine levels are lowered (by giving animals a competing neutral amino acid) (*Wurtman et al.*, 1974). The consumption of individual meals that change brain tyrosine levels also causes the expected changes in brain dopa accumulation: a meal that is relatively rich in protein causes disproportionate increases in plasma tyrosine (because some of the phenylalanine in the protein is converted to tyrosine in the liver), thereby increasing both brain tyrosine levels and the rate of catechol synthesis. Available evidence suggests that there are some differences between dopaminergic and noradrenergic brain neurons in the extent to which production of their neurotransmitters depends on precursor (tyrosine) levels: In dopaminergic neurons, the tyrosine hydroxylase must be activated (*e.g.*, by treating animals with a drug, haloperidol, that accelerates the neurons' firing rates) in order for tyrosine levels to control dopamine formation (*Scally et al.*, 1977); in contrast, in all situations thus far examined brain norepinephrine synthesis appears to be precursor-dependent (*Gibson* and *Wurtman*, 1978). Precursor-induced changes in catecholamine synthesis *are* associated with parallel changes in the release of dopamine and norepinephrine (as indicated by changes in brain levels of homovanillic acid and MOPEG sulphate, their major metabolites) (*Scally et al.*, 1977; *Gibson* and *Wurtman*, 1977).

The ability of brain tyrosine levels to affect catecholamine synthesis could have an even greater impact on clinical problems than the uses described above for choline's effects on acetylcholine. Norepinephrine-containing neurons are involved in a very large number of physiological mechanisms, both inside and outside the brain, and dopaminergic neurons have clearly been implicated in the etiology of Parkinson's disease, schizophrenia, and other brain disorders.

Acknowledgements

The studies described in this report have been supported in part by grants from the United States Public Health Service.

References

Cohen, E. L., Wurtman, R. J.: Brain acetylcholine: increase after systemic choline administration. Life Sciences *16*, 1095—1102 (1975).

Cohen, E. L., Wurtman, R. J.: Brain acetylcholine synthesis: control by dietary choline. Science *191*, 561—562 (1976).

Davis, K. L., Berger, P. A., Hollister, L. E.: Choline for tardive dyskinesia. New Eng. J. Med. *293*, 152 (1975).

Fernstrom, J. D., Wurtman, R. J.: Brain serotonin content: physiological dependence on plasma tryptophan levels. Science *173*, 149—152 (1971 a).

Fernstrom, J. D., Wurtman, R. J.: Brain serotonin content: increase following ingestion of carbohydrate diet. Science *174*, 1023—1025 (1971 b).

Fernstrom, J. D., Wurtman, R. J.: Elevation of plasma tryptophan by insulin in the rat. Metabolism *21*, 337—342 (1972 a).

Fernstrom, J. D., Wurtman, R. J.: Brain serotonin content: regulation by plasma neutral amino acids. Science *178*, 414—416 (1972 b).

Fernstrom, J. D., Wurtman, R. J.: Nutrition and the brain. Scientific American *230*, 84—91 (1974).

Gibson, C. J., Wurtman, R. J.: Physiological control of brain norepinephrine synthesis by brain tyrosine concentration. Life Sciences *22*, 1399—1406 (1978).

Growdon, J. H., Hirsch, M. J., Wurtman, R. J., Weiner, W.: Oral choline administration to patients with tardive dyskinesia. New Eng. J. Med. *297*, 524—527 (1977).

Growdon, J. H., Gelenberg, A. J., Doller, L., Hirsch, M. J., Wurtman, R. J.: Lecithin can suppress tardive dyskinesia. New Eng. J. Med. *298*, 1029 to 1030 (1978).

Hirsch, M. J., Growdon, J. H., Wurtman, R. J.: Relations between dietary choline intake, serum choline levels, and various metabolic indices. Metabolism *27*, 953—960 (1978).

Pardridge, W. M.: Regulation of amino acid availability to the brain. In: Nutrition and the Brain, Vol. I (*Wurtman, R. J., Wurtman, J. J.*, eds.), pp. 141—204. New York: Raven Press. 1977.

Scally, M. J., Wurtman, R. J.: Brain tyrosine level controls striatal dopamine synthesis in haloperidol-treated rats. J. Neural Transm. *41*, 1—6 (1977).

Ulus, I., Scally, M. J., Wurtman, R. J.: Enhancement by choline of the induction of adrenal tyrosine hydroxylase by phenoxybenzamine, 6-hydroxydopamine, insulin, or exposure to cold. J. Pharmacol. Exp. Ther. *204*, 676—682 (1978).

Wurtman, J. J., Wurtman, R. J.: Fenfluramine and fluoxetine spare protein consumption while suppressing caloric intake by rats. Science *198*, 1178—1180 (1977).
Wurtman, R. J., Larin, F., Mostafapour, S., Fernstrom, J. D.: Brain catechol synthesis: control by brain tyrosine concentration. Science *185*, 183—184 (1974).
Wurtman, R. J., Fernstrom, J. D.: Control of brain neurotransmitter synthesis by precursor availability and nutritional state. Biochem. Pharmacol. *25*, 1691—1696 (1976).
Wurtman, R. J., Hirsch, M. J., Growdon, J. H.: Lecithin consumption raises serum free choline levels. Lancet *II*, 68—69 (1977).

Author's address: Prof. *R. J. Wurtman,* Laboratory of Neuroendocrine Regulation, Department of Nutrition and Food Science, Cambridge, MA 02139, U.S.A.

Journal of Neural Transmission, Suppl. 15, 81—92 (1979)

Relationships Between Plasma, CSF and Brain Tryptophan

G. Curzon

Department of Neurochemistry, Institute of Neurology, London, U.K.

With 3 Figures

Summary

In many circumstances plasma free tryptophan correlated better than plasma total tryptophan with brain tryptophan concentration (immobilization, fasting, acute liver failure, some drugs). Also, using a modified Oldendorf method it was found that changes of plasma tryptophan binding considerably affected brain tryptophan uptake. Usually, changes of plasma tryptophan binding and non-esterified fatty acid concentration were associated. This led either to changes of plasma free and brain tryptophan concentrations (see above) or to "buffering" in which the proportion of plasma tryptophan in the free state changed but not its concentration.

The plasma free tryptophan-brain tryptophan relationship was confirmed in rats after portocaval anastomosis or sham operation. In these experiments brain tryptophan changes did not correlate with plasma amino acids competing with tryptophan for transport to the brain.

Determinations on plasma, lumbar and ventricular CSF from psychiatric patients suggest that plasma free tryptophan concentration provides an index of CSF tryptophan and 5-hydroxyindoleacetic acid concentrations and hence of 5-hydroxytryptamine turnover in the human central nervous system.

Neurochemists study relationships between tryptophan concentrations in plasma, CSF and brain for two related reasons. Firstly these relationships may be used to elucidate *mechanisms* by which brain tryptophan concentrations is influenced. Secondly, tryptophan concentration in more accessible compartments (*i.e.* plasma, CSF) may be used as an index of its concentration in the brain and hence of its availability for 5-HT synthesis therein. I will be discussing work in my laboratory on relationships between rat plasma and brain tryptophan concentrations in various circumstances and will then describe some work on the predictive value of plasma tryptophan measurements with respect to tryptophan metabolism in the human central nervous system.

Relationships Between Plasma and Brain Tryptophan Concentrations in the Rat

Although brain and plasma tryptophan concentrations are significantly and positively correlated in some experiments this is by no means an invariable finding. Thus, results in Table 1 show a highly significant correlation between plasma and brain tryptophan of untreated rats in Experiment 1 but in two other experiments in which the rats were kept in a different and less quiet animal housing significant correlations were not obtained for untreated, immobilized or food deprived rats. However, brain-plasma correlations for another aromatic amino acid, tyrosine were on the whole significant and positive.

Plasma Tryptophan Binding

Much of our work has derived from this frequent finding of the lack of a simple relationship between plasma and brain tryptophan.

Table 1. *Relationships between plasma and brain tryptophan and tyrosine concentrations within groups of rats*

Group	Correlation Coefficient			
	Plasma v. Brain			
	Tryptophan		Tyrosine	
Experiment 1				
Controls (24)	0.53	$P < 0.01$	0.64	$P < 0.001$
Experiment 2				
Controls (11)	0.14	ns	0.62	$P < 0.05$
Immobilized (11)	—0.37	ns	0.80	$P < 0.01$
Experiment 3				
Controls (23)	0.33	ns	0.74	$P < 0.001$
Food deprived 24 hours (24)	—0.19	ns	0.41	ns

Rats were male Sprague-Dawley, 100—200 g (Anglia Laboratory Animals, Alconbury, Cambs.). Experiments 2 and 3 are from *Curzon, Joseph,* and *Knott* (1972). Experiment 1 was performed using identical methods except that the rats were taken from a quieter housing (*Kantamaneni, Fernando,* and *Curzon,* unpublished work).

Initially, the fact that tryptophan (alone among the plasma amino acids) was largely bound to albumin suggested that there might be a unique restriction on the transport of tryptophan to the brain (*Knott* and *Curzon,* 1972). Subsequent work in our own and in other laboratories showed that plasma free tryptophan concentration (determined after separation by ultrafiltration or dialysis methods) was a better predictor of brain tryptophan concentration than was plasma total tryptophan (determined after protein precipitation). For example, immobilization, fasting (*Knott* and *Curzon,* 1972), acute liver failure either obtained by portocaval anastomosis (*Curzon, Kantamaneni, Winch, Rojas-Bueno, Murray-Lyon,* and *Williams,* 1973; *Cummings, James, Soeters, Keane, Foster,* and *Fischer,* 1976) or by carbon tetrachloride poisoning (*Knott* and *Curzon,* 1975) and various drug treatments (*Curzon* and *Knott,* 1974; *Gessa* and *Tagliamonte,* 1974) all increased plasma free but not total tryptophan and also increased brain tryptophan.

In most of the above studies the rise of plasma free tryptophan was associated with a rise of plasma non-esterified fatty acid (NEFA) which weakens the binding of tryptophan to albumin by itself becoming bound to it (*McMenamy,* 1965). While in most of these experiments plasma total tryptophan concentration showed relatively little change, under other circumstances an increase of plasma NEFA is associated with a marked and rapid fall of plasma total tryptophan *e.g.* when rats previously fasted for 24 hours are disturbed by removal from group housing and caging separately (*Knott, Hutson,* and *Curzon,* 1977). In these experiments (Table 2) the *fraction* of plasma tryptophan in the free state rose but concurrently plasma total tryptophan fell. Therefore, the rise of free tryptophan *concentration* was relatively small, and in some experiments was negligible. How-

Table 2. *Effect of 8 min cage disturbance on plasma and brain tryptophan*

	Percentage changes from controls		
	Plasma tryptophan		Brain
Experiment	Total	Free	Tryptophan
1	—39 ($P < 0.001$)	+2	—2
2	—56 ($P < 0.001$)	+40 ($P < 0.05$)	+3
3	—23 ($P < 0.05$)	+80 ($P < 0.05$)	+36 ($P < 0.05$)
4	—14	+80 ($P < 0.001$)	+24 ($P < 0.05$)

Rats were caged in groups of 3, deprived of food for 24 hours, and transferred to individual cages under somewhat different conditions for each experiment as described by *Knott, Hutson,* and *Curzon* (1977).

ever, in those experiments in which plasma free tryptophan concentration rose considerably then so did brain tryptophan albeit to a smaller extent. When plasma free tryptophan concentration altered less markedly then brain tryptophan showed little change. Similarily, in more prolonged experiments (see below) in which plasma tryptophan was displaced from albumin by injection of tolbutamide a negligible change of plasma free tryptophan concentration together with a striking fall of total tryptophan were associated with a negligible change of brain tryptophan concentration.

Thus, the release of tryptophan from plasma albumin appears to lead to increased plasma free and brain tryptophan concentrations in some situations but to a kind of buffering effect in others so that both plasma free and brain tryptophan levels remain essentially unaltered. It may be that the fall of plasma total tryptophan also has some influence on brain tryptophan concentration (*Yuwiler, Oldendorf, Geller,* and *Braun,* 1977) but this is not clearly indicated by our results. What determines which of the above two consequences of plasma tryptophan release occurs is of obvious importance but remains obscure.

Insulin and Competing Amino Acids

The results I have described indicate that prediction of brain tryptophan concentration is helped by information on plasma free tryptophan concentration. They do not necessarily imply that plasma free tryptophan is the sole influence or even the most important influence on brain tryptophan. Indeed, they do not necessarily imply that brain tryptophan is at all influenced by plasma free tryptophan —only that its concentration can be to some extent predicted from that of plasma free tryptophan. In certain circumstances other factors may have dominant effects on brain tryptophan. Thus it is well established that large change of insulin secretion such as occur on food intake (*Fernstrom* and *Wurtman,* 1971) or on production of experimental diabetes (*Fernando, Knott,* and *Curzon,* 1976; *Mackenzie* and *Trulson,* 1978) may affect brain tryptophan and that the mechanism responsible involves altered plasma concentrations of amino acids which compete with tryptophan for uptake by the brain. Nevertheless, a number of drugs which cause moderate or even quite large changes of insulin secretion (propranolol, tolbutamide, phentolamine) do not appear to affect brain uptake of tryptophan or other aromatic amino acids insofar as this is indicated by their brain/plasma ratios (*Curzon* and *Fernando,* 1977).

A scheme incorporating the apparent influences of both plasma tryptophan binding and insulin secretion on brain tryptophan was suggested by a study of the effect of aminophylline on brain tryptophan (*Curzon* and *Fernando*, 1976). At moderate dosage this drug has a lipolytic effect so that plasma free tryptophan rises and there is an associated rise of brain tryptophan. At higher doses of aminophylline the increases of brain tryptophan are no longer explicable by changes of plasma free tryptophan and are presumably related to the concurrent very high plasma insulin values. Thus, if aminophylline at higher dosage is given to streptozotocin diabetic rats then insulin does not rise and the increase of brain tryptophan is proportionate to that of plasma free tryptophan. These results conform to the scheme shown in Fig. 1 which indicates how supplies of tryptophan to the brain may be maintained in two contrasting dietary situations—as a result of insulin secretion after food intake;—as a result of increased plasma NEFA (UFA) after moderate periods of food deprivation.

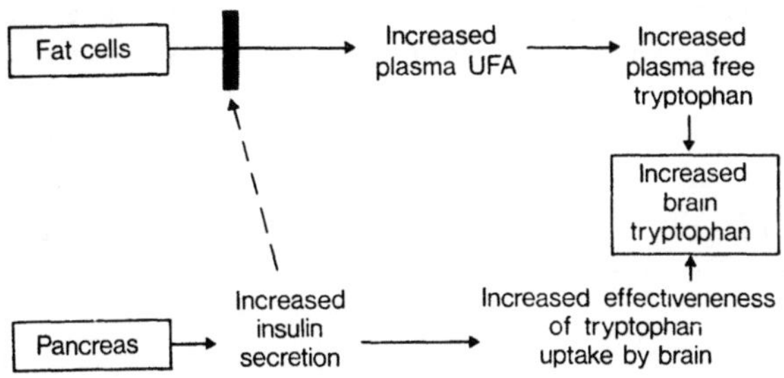

Fig. 1. Effect of lipolysis and insulin secretion on brain tryptophan (modified from *Curzon* and *Fernando*, 1976)

Despite many findings which are difficult to explain otherwise, doubt has been cast on whether any relationship exists between plasma free tryptophan and brain tryptophan. Some apparently discrepant results which have been reported may have been due to artefactual changes—in particular to the extremely active *in vitro* lipolysis in some lipaemic plasmas which can lead to falsely high free tryptophan values (*Knott*, *Hutson*, and *Curzon*, 1976). A more serious source of uncertainty is that plasma competing amino acids have been determined in few of the experiments in which plasma free tryptophan appears to influence brain tryptophan.

Therefore the following results obtained during an investigation of the increase of rat brain tryptophan after portocaval anastomosis (*Bloxam* and *Curzon*, 1978) may be of interest. Regression analysis of brain tryptophan concentrations against plasma total and free tryptophan concentrations obtained at different times after operation (Fig. 2) reveals that correlations between brain and plasma total tryptophan are significant for neither sham operated nor anastomosed animals but that correlations between brain and plasma free tryptophan are significant for both groups. Similar correlations were still indicated when the influence of amino acids competing with tryptophan for uptake by brain was corrected for. This was done using the procedure of *Pardridge* (1977) in which it is assumed that tryptophan is largely transported to the brain by a mechanism for which large neutral amino acids compete with K_i values equivalent to their K_m values for uptake.

Recently we have done some experiments on rat brain tryptophan uptake using the Oldendorf method (Table 3). With a procedure in which uptake was measured from a bolus of pH 7.4 Krebs-Ringer buffer ±3 % bovine albumin injected into the carotid our results essentially agree with those of *Yuwiler et al.* (1977) as the absence of albumin from the bolus only increased uptake by 37 % though free tryptophan concentration rose three-fold. However under more physiological conditions *i.e.* when uptake was measured not from buffer but from the animals own reinjected plasma the uptake was increased by 52 % when free tryptophan was doubled by addition of

Table 3. *Plasma tryptophan binding and tryptophan uptake by brain*

Bolus injected	Tryptophan in bolus nmole/ml Total	Free	Brain Tryptophan Uptake
Experiment 1			
Buffer + albumin (10)	69	23	1.91 ± 0.24
Buffer (13)	69	69	2.62 ± 0.22
Buffer + clofibrate (8)	69	69	2.74 ± 0.18*
Experiment 2			
Plasma (12)	80 ± 21	24 ± 7	0.85 ± 0.22
Plasma + clofibrate (9)	76 ± 19	49 ± 11	1.29 ± 0.26*

Brain tryptophan uptake from an intracarotid bolus was measured as *Yuwiler et al.* (1977) using L-methylene ^{14}C-tryptophan + ^{3}H-water. Results shown ± S.D. (n) as brain uptake index X bolus tryptophan concentration (nmole/ml). In Experiment 1, the bolus contained Krebs-Ringer phosphate buffer pH 7.4 and bovine albumin 3 % or clofibrate 0.37 mM. In Experiment 2, 1.0 ml blood was withdrawn from each rat and 0.2 ml of the plasma reinjected after adding L-methylene ^{14}C-tryptophan + ^{3}H-water with or without clofibrate. Differences from controls without clofibrate: * $P < 0.001$.

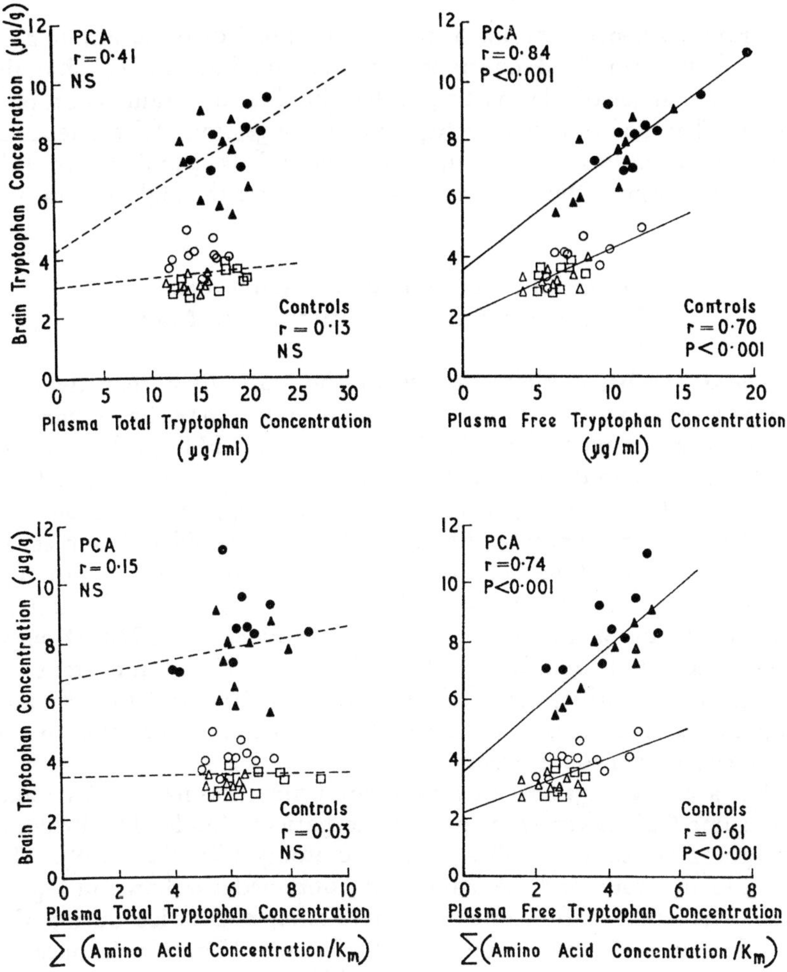

Fig. 2. Relationship between brain tryptophan concentration and *a* plasma total tryptophan concentration, *b* plasma free tryptophan concentration, *c* plasma total tryptophan concentration divided by $\sum \frac{[s]}{K_m}$, *d* plasma free tryptophan concentration divided by $\sum \frac{[s]}{K_m}$.

$\sum \frac{[s]}{K_m}$ is the sum of the plasma concentrations of valine, leucine, isoleucine, tyrosine and phenylalanine, each divided by the corresponding K_m value and represents their combined competitive inhibitory effects on tryptophan uptake. Kinetics of this type *in vivo* are indicated by *Pardridge* (1977). Results are from 20 rats killed 1 day (●) or 10 days (▲) after portocaval anastomosis (PCA) and 29 rats killed unoperated (□) or 1 day (○) or 10 days (△) after sham operation. Lines were obtained by linear regression. Results from *Bloxam* and *Curzon* (1978)

clofibrate. As concentrations of total tryptophan or other amino acids in the bolus remained constant these results indicate considerable effects of changes of plasma tryptophan binding on brain tryptophan uptake. They do not however disprove the suggestion of *Yuwiler et al.* that the brain has some ability to strip off tryptophan from its binding to plasma albumin when the blood passes through the brain.

Relationships Between Plasma and CSF Tryptophan Concentrations in Man

There has been much interest in the possibility that deficient transport of tryptophan to the human brain may have harmful consequences. In particular, the defect in brain 5-HT synthesis which may have a role in depressive illness has been suggested to result from low plasma free tryptophan concentrations (*Coppen, Eccleston,* and *Peet,* 1973). Unfortunately, direct attack on this problem is subject to many difficulties. One serious problem is that brain tryptophan concentration rises considerably after death and possibly also during terminal stages of life (*Curzon* and *Kantamaneni,* 1978; *Young, Tsang, Lal,* and *Sourkes,* 1977). Therefore it is important to investigate the utility of plasma total and free tryptophan and competing amino acids as indices of tryptophan concentration in the human central nervous system.

We are investigating this problem using plasma, lumbar and ventricular CSF taken from psychiatric patients (under Dr. Bridges) undergoing stereotactic subcaudate tractotomy (*Knight,* 1969). The scheme of the study and the main results obtained are shown in Fig. 3. Blood was taken before operation at 8 a.m., and twice during the operation at 8.45 a.m.—9 a.m. and 9.45 a.m.—10 a.m. and plasmas separated. Ventricular and lumbar fluids were taken concurrently with the second and third blood samples respectively. A number of significant correlations were found:

(1) There was a clear difference between plasmas taken before and during anaesthesia. Before anaesthesia, plasma total tryptophan showed a significant negative correlation with NEFA concentration so that although the proportion of plasma tryptophan in the free state correlated significantly with NEFA the free tryptophan concentration did not. During anaesthesia, plasma total tryptophan was significantly lower but within the group of plasmas collected at each time the variation of NEFA concentration was not associated with plasma total tryptophan differences. Instead, highly significant positive correlations were apparent between NEFA and free tryptophan con-

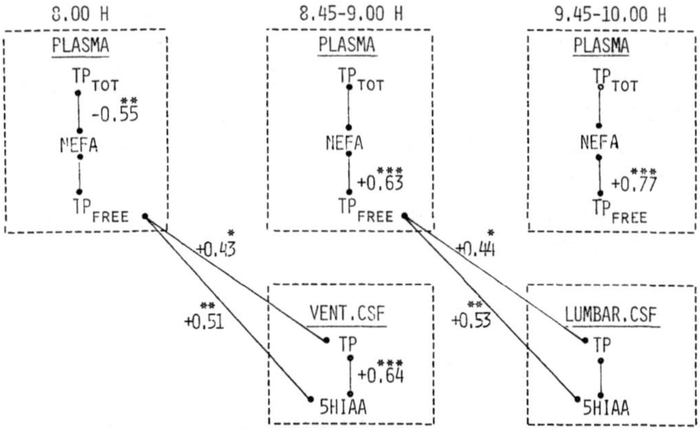

Fig. 3. Relationships between concentrations of non-esterified fatty acid (NEFA) and indolic substances in plasma, ventricular CSF and lumbar CSF of psychiatric subjects undergoing stereotactic tractotomy. Samples were collected at times indicated. Anaesthetic treatment began at 8.30 a.m. (barbiturate i.v. then halothane). Correlation coefficients between concentrations of different substances are shown *(when statistically significant)* * $P < 0.05$, ** $P < 0.02$, *** $P < 0.01$, $n = 19$—22. (*Erratum:* The correlation coefficient between the concentration of TP_{TOT} and NEFA at 8 a.m. should be —0.65*** instead of —0.55**)

centration. Thus, samples before anaesthesia show plasma NEFA changes leading to a buffer-like action on plasma free tryptophan while samples during anaesthesia show plasma NEFA changes leading to changes of plasma free tryptophan concentration.

(2) In confirmation of our previous results (*Curzon, Kantamaneni, Bartlett,* and *Bridges,* 1976) ventricular CSF tryptophan and 5-HIAA concentrations showed a highly significant correlation. This is consistent with tryptophan availability having an important influence on brain 5-HT turnover in man. However, a similar relationship was not apparent for the lumbar CSF.

(3) No statistically significant relationships were found between concentrations in plasmas and CSF samples collected at the same time but plasma free tryptophan (though not total tryptophan) concentrations correlated at the $<.05$ level with tryptophan concentrations in ventricular or lumbar CSF collected 1 hour later. Corresponding correlations between plasma free tryptophan and CSF 5-HIAA were rather more significant.

The evidence for a relationship between human plasma free and CSF tryptophan concentrations is consistent with work by *Young, Lal, Feldmuller, Sourkes, Ford, Kiely,* and *Martin* (1976) in which tryptophan was determined in plasma and CSF samples taken during ventricular drainage and with our own results on material from control and uraemic patients (*Sullivan, Murnaghan, Callaghan, Kantamaneni*, and *Curzon*, 1978).

Results in general therefore point to plasma free tryptophan values as moderately useful indices of CSF tryptophan and 5-HIAA concentration and hence of 5-hydroxytryptamine turnover in the human central nervous system. However there is evidence that competing amino acids may also have important relevant effects (*Perez-Cruet, Chase*, and *Murphy*, 1974; *Møller, Kirk*, and *Fremming*, 1976).

Acknowledgement

The Medical Research Council is thanked for financial support.

References

Bloxam, D. L., Curzon, G.: A study of proposed determinants of brain tryptophan concentration in rats after portocaval anastomosis or sham operation. J. Neurochem. *31*, 1255—1263 (1978).

Coppen, A., Eccleston, E. G., Peet, M.: Total and free tryptophan concentration in the plasma of depressive patients. Lancet *ii*, 60—63 (1973).

Curzon, G., Joseph, M. H., Knott, P. J.: Effects of immobilization and food deprivation on rat brain tryptophan metabolism. J. Neurochem. *19*, 1967—1974 (1972).

Curzon, G., Kantamaneni, B. D., Winch, A., Rojas-Bueno, I., Murray-Lyon, M., Williams, R.: Plasma and brain tryptophan changes in experimental acute hepatic failure. J. Neurochem. *21*, 137—146 (1973).

Curzon, G., Knott, P. J.: Effects on plasma and brain tryptophan in the rat of drugs and hormones that influence the concentration of unesterified fatty acid in the plasma. Br. J. Pharmac. *50*, 197—204 (1974).

Curzon, G., Fernando, J. C. R.: Effect of aminophylline on tryptophan and other aromatic amino acids in plasma, brain and other tissues and on brain 5-hydroxytryptamine metabolism. Br. J. Pharmac. *58*, 533—545 (1976).

Curzon, G., Kantamaneni, B. D., Bartlett, J. R., Bridges, P. K.: Transmitter precursors and metabolites in human ventricular cerebrospinal fluid. J. Neurochem *26*, 613—615 (1976).

Curzon, G., Fernando, J. C. R.: Drugs altering insulin secretion: effects on plasma and brain concentrations of aromatic amino acids and on brain 5-hydroxytryptamine turnover. Br. J. Pharmac. *60*, 401—408 (1977).

Curzon, G., Kantamaneni, B. D.: Transmitters and related substances in brain material: an index of terminal and/or post-mortem biochemical change. Lancet *i*, 156—157 (1978).

Cummings, M. G., James, J. H., Soeters, P. B., Keane, J. M., Foster, J., Fischer, J. E.: Regional brain study of indoleamine metabolism in the rat in acute hepatic failure. J. Neurochem. *27,* 741—746 (1976).

Fernando, J. C. R., Knott, P. J., Curzon, G.: The relevance of *both* plasma free tryptophan and insulin to rat brain tryptophan concentration. J. Neurochem. *27,* 343—345 (1976).

Fernstrom, J. D., Wurtman, R. J.: Brain serotonin content: increase following ingestion of carbohydrate diet. Science *174,* 1023—1025 (1971).

Gessa, G. L., Tagliamonte, A.: Possible role of free serum tryptophan in the control of brain tryptophan concentration and serotonin synthesis. Adv. Biochem. Psychopharmac. *11,* 119—132 (1974).

Hutson, P. H., Knott, P. J., Curzon, G.: Control of brain tryptophan concentration in rats on a high fat diet. Nature *262,* 142—143 (1976).

Knight, G. C.: Bi-frontal stereotactic tractotomy: an atraumatic operation of value in the treatment of intractable psychoneurosis. Brit. J. Psychiat. *115,* 257—266 (1969).

Knott, P. J., Curzon, G.: Free tryptophan in plasma and brain tryptophan metabolism. Nature *239,* 452—453 (1972).

Knott, P. J.: Curzon, G.: Tryptophan and tyrosine disposition and brain tryptophan metabolism in acute carbon tetrachloride poisoning. Biochem. Pharmacol. *24,* 963—966 (1975).

Knott, P. J., Hutson, P. H., Curzon, G.: Fatty acid and tryptophan changes on disturbing groups of rats and caging them singly. Pharmacol. Biochem. Behav. *7,* 245—252 (1977).

Mac Kenzie, R. G., Trulson, M. E.: Effects of insulin and streptozotocin-induced diabetes on brain tryptophan and serotonin metabolism in rats. J. Neurochem. *30,* 205—211 (1978).

McMenamy, R. H.: Binding of indole analogues to human serum albumin. Effects of fatty acids. J. biol. Chem. *24,* 4235—4243 (1965).

Møller, S. E., Kirk, L., Fremming, K. H.: Plasma amino acids as an index for subgroups in manic depressive psychosis: correlation to effect of tryptophan. Psychopharmacology *49,* 205—213 (1976).

Pardridge, W. M.: Kinetics of competitive inhibition of neutral amino acid transport across the blood-brain barrier. J. Neurochem. *28,* 103—108 (1977).

Perez-Cruet, J., Chase, T. N., Murphy, D. L.: Dietary regulation of brain tryptophan metabolism by plasma ratio of free tryptophan and neutral amino acids in humans. Nature *248,* 693—695 (1974).

Sullivan, P. A., Murnaghan, D., Callaghan, N., Kantamaneni, B. D., Curzon, G.: Cerebral transmitter precursors and metabolites in advanced renal disease. J. Neurol. Neurosurg. Psychiat. *41,* 581—588 (1978).

Young, S. N., Lal, S., Feldmuller, F., Sourkes, T. L., Ford, R. M., Kiely, M., Martin, J. B.: Parallel variation of ventricular CSF tryptophan and free serum tryptophan in man. J. Neurol. Neurosurg. Psychiat. *39,* 61—65 (1976).

Young, S. N., Tsang, D., Lal., S., Sourkes, T. L.: Changes in the tryptophan content of excised human cerebral cortex. J. Neurochem. *28*, 439—440 (1977).

Yuwiler, A., Oldendorf, W. H., Geller, E., Braun, L.: Effect of albumin binding and amino acid competition on tryptophan uptake into brain. J. Neurochem. *27*, 1015—1023 (1977).

Author's address: Dr. *G. Curzon*, Department of Neurochemistry, Institute of Neurology Laboratories, 33 John's Mews, London WC1N 3NS, U.K.

Journal of Neural Transmission, Suppl. 15, 93—105 (1979)

Ontogenesis of Tryptophan Transport in the Rat Brain

M. Hamon and **S. Bourgoin**

Groupe NB-INSERM U 114, Collège de France, Paris, France

With 4 Figures

Summary

During the first three postnatal weeks, the levels of tryptophan in the brain are exceptionally high, 2—4 times those found in adult rats. This is related to two main peculiarities concerning tryptophan transport in young animals: 1. the lack of tryptophan binding onto serum albumin which makes its diffusion from plasma to tissues easier for the early life period; 2. the greater capacity of synaptosomes from neonates to accumulate tryptophan.

Experiments consisting of electrolytic lesioning of the midbrain raphé or 5, 7-dihydroxytryptamine treatment clearly demonstrate that the uptake of tryptophan during postnatal development is not more active in serotoninergic than in other types of nerve terminals.

In adult rats, changing the concentration of tryptophan induces parallel modifications in the rate of 5-HT synthesis in the brain since the rate limiting enzyme, tryptophan hydroxylase, is not saturated by its substrate. In contrast, neither tryptophan loading, nor parachlorophenylalanine administration (resulting in a marked decrease in brain tryptophan levels) alters the rate of 5-HT synthesis in the CNS of neonates, indicating that tryptophan hydroxylase is saturated during the early life period. These results are discussed in relation to the possible non-transmitter role of 5-HT during brain growth.

In mammals, tryptophan (TRP) is a unique amino acid for several reasons: 1. it is the only amino acid which is bound to serum albumin in plasma of adult animals (see *Curzon* and *Knott*, 1977; *Gessa* and *Tagliamonte*, 1974; *Hamon* and *Glowinski*, 1974); 2. it is the precursor of serotonin (5-HT), one putative neurotransmitter in the central nervous system (CNS); 3. under some particular circumstances, its concentration can be the rate limiting factor for the biosynthesis of proteins in the periphery (*Sidransky et al.*, 1968) as well as in the brain during ontogenesis (see *Barondes*, 1974).

Since tryptophan is an essential amino acid in mammals, its supply to the cell machinery, notably for the synthesis of proteins and 5-HT, is entirely carried out by specific transport mechanisms from the blood circulation. In adult rats, they involve the binding of tryptophan onto serum albumin and an amino acid carrier in the cell membrane. *In vivo* (*De Montis et al.*, 1977; *Etienne et al.*, 1976) as well as *in vitro* (*Korpi* and *Oja*, 1978), the binding of tryptophan onto serum albumin limits its capacity to enter the cells, including serotoninergic neurons. Therefore, treatments known to increase the concentration of free tryptophan in serum generally induce both an enhanced accumulation of the amino acid and an acceleration of 5-HT synthesis in the CNS of adult rats (*Bourgoin et al.*, 1974; *Curzon* and *Knott*, 1977; *Gessa* and *Tagliamonte*, 1974; *Knott et al.*, 1973). Under physiological conditions, neutral amino acids in plasma compete for the tryptophan carrier in membranes so that alterations in their levels can also affect the transport of tryptophan from blood circulation to the brain (*Fernstrom* and *Wurtman*, 1972; *Pérez-Cruet et al.*, 1974). Finally, the last factor which controls the concentration of tryptophan in cells is the amino acid carrier itself. Changes in its intrinsic properties have already been shown to result in parallel modifications in the rate of 5-HT synthesis *in vivo* as well as *in vitro* (*Hamon et al.*, 1974; *Hamon* and *Glowinski*, 1974).

Since the levels of tryptophan in brain tissues are exceptionally high in newborn animals (*Bourgoin et al.*, 1974; *Hoff et al.*, 1974; *Miller et al.*, 1977; *Tyce et al.*, 1964), this might result from peculiarities concerning its transport during the early life period. In this short review, we describe the characteristics of the transport of tryptophan in blood and those exhibited by the amino acid carrier in synaptosomes from very young rats. Both mechanisms are perfectly adapted to supporting a high level of protein synthesis in the growing brain. In addition, they lead to the synthesis of 5-HT at a maximum rate for a few days following birth. In view of the proposed non-transmitter role of 5-HT during development (see *Baker* and *Quay*, 1969), this transient acceleration of its synthesis might be of critical importance.

1. Characteristics of Tryptophan Binding in Serum During Development in the Rat

As shown in Fig. 1, tryptophan levels in the CNS are much higher immediately after birth than during the adult life. They decrease rapidly for the first postnatal week and more slowly thereafter to reach adult values at the end of the 4th week after birth. The comparison of these developmental changes with those of circulating tryptophan (Fig. 2) led to striking findings: whereas there was no correlation between the levels of the amino acid in the brain and the concentration of whole tryptophan in serum, the time-course of changes in the concentration of circulating free tryptophan closely resembled that found in tissues. Thus, at birth, when brain levels of the amino acid were maximum (Fig. 1), the concentration of free tryptophan was so high that in fact no bound tryptophan was detected in serum (Fig. 2). Later, a bound fraction appeared and increased progressively reducing the proportion of diffusible free tryptophan to less than 10 % of the total amount of circulating amino acid (Fig. 2). These findings first reported by *Bourgoin et al.* (1974) have been recently confirmed by *Miller et al.* (1977).

According to *Bourgoin et al.* (1976, 1977 b), several factors account for the poor binding of tryptophan during the early life period: 1. the concentration of the binding protein, serum albumin,

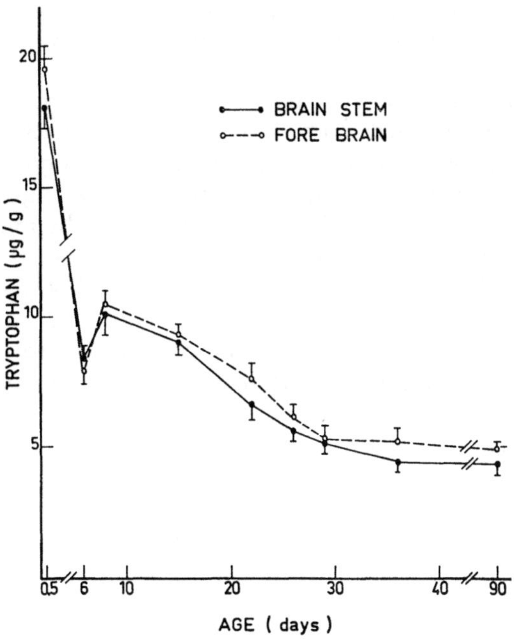

Fig. 1. Developmental changes in tryptophan levels in the brain stem and fore brain of the rat. Animals (Sprague-Dawley, Charles River strain) were killed at various times after birth (0 in the abscissa). Tryptophan was measured as described by *Bourgoin et al.* (1974). Each point is the mean ± SEM of 6—8 determinations

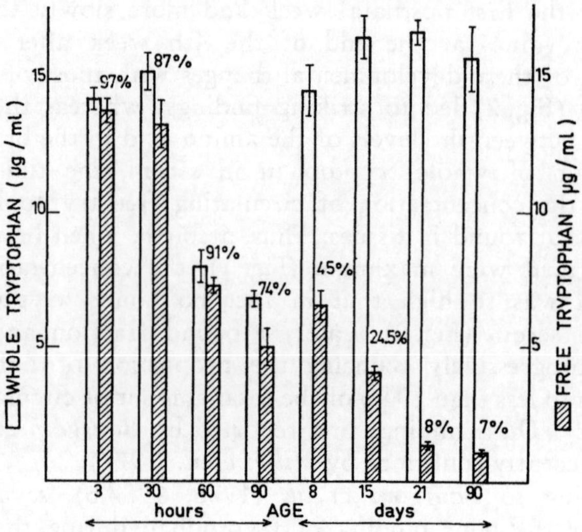

Fig. 2. Tryptophan concentrations in rat serum at various postnatal ages. The empty bars represent total tryptophan and the hatched ones, the free form of the amino acid measured as described by *Bourgoin et al.* (1974). Percentage on bars express the ratio of the concentration of free tryptophan over that of total amino acid. Each value is the mean ± SEM of 8—10 determinations

remains significantly lower than in adults for the three postnatal weeks; 2. non-esterified fatty acids which are natural competitive inhibitors of tryptophan binding are relatively abundant in the serum for the whole period of lactation (as a result of the high fat milk diet); 3. the intrinsic properties of the defatted serum albumin purified from the serum of newborn rats are different from those of the protein from adult animals. Whereas the total number of binding sites per molecule of serum albumin was very close to one in adult animals, the reduced capacity of the defatted protein from neonates seemed to be related to a corresponding value significantly lower than one. However, immunological studies failed to reveal any structural difference between the adult and the neonatal protein (*Bourgoin et al.*, 1977 b).

Among drugs known to alter the rate of 5-HT synthesis in adults, several of them act at the periphery by changing the proportion of free tryptophan (*Curzon* and *Knott*, 1977; *Gessa* and *Tagliamonte*, 1974; *Hamon* and *Glowinski*, 1974). For instance, salicylate first displaces tryptophan from its binding site onto serum albumin, causing an increased accumulation of the amino acid and an acceleration of 5-HT synthesis in the CNS (*Bourgoin et al.*, 1974; *Gessa* and *Tagliamonte*, 1974). In contrast, treatment with this drug during the early life period fails to increase the rate of central 5-HT synthesis since tryptophan is already entirely free in serum (*Bourgoin et al.*, 1974). Similar findings have been obtained with chlordiazepoxide, another competitive inhibitor of tryptophan binding onto serum albumin (*Bourgoin et al.*, 1974; *Hamon* and *Glowinski*, 1974).

In normal rats, the failure of tryptophan-displacing drugs to alter the synthesis of 5-HT in brain only concerns a very short period after birth, since a significant fraction of circulating tryptophan is bound to serum albumin as soon as the 4—5th postnatal day (Fig. 2, *Bourgoin et al.*, 1974; *Miller et al.*, 1977). Recently, we observed that intrauterine growth retardation induced a significant delay in the appearance of bound tryptophan in serum. Indeed, in 8 day-old hypotrophic rats (obtained from a dam whose uterine artery was ligatured on the 17th day of gestation according to *Wigglesworth* 1964), circulating tryptophan was still entirely free (Table 1). In the CNS, this resulted in tryptophan and 5-hydroxyindole acetic acid (5-HIAA) levels in hypotrophic rats markedly higher than in normal pair-aged controls (Table 1). Further studies revealed that, in contrast to what occurred in normal rats (Fig. 1), brain tryptophan levels did not decrease for the first two weeks after birth because of the poor binding of serum tryptophan in hypotrophic rats (*Chanez-Bel et al.*,

Table 1. *Effect of intrauterine growth retardation on the binding of TRP in serum and on the levels of TRP, 5-HT and 5-HIAA in the forebrain on the 8th postnatal day*

		Control	Hypotrophic
Serum	TYR	56.69 ± 2.88	53.04 ± 3.25
	Whole TRP	32.44 ± 2.88	22.41 ± 2.74*
	Free TRP	14.46 ± 1.12	20.18 ± 1.05*
	%	44.6	90.0
Forebrain	TRP	8.74 ± 0.78	12.83 ± 1.23*
	5-HT	0.25 ± 0.01	0.27 ± 0.03
	5-HIAA	0.49 ± 0.02	0.68 ± 0.04*

Intrauterine growth retardation was achieved by the ligation of one uterine artery at its proximal end on the seventeenth day of gestation (*Wigglesworth*, 1964). After delivery, hypotrophic and control newborn rats were reared under similar conditions with a lactating female fed *ad libitum*. Young animals were killed on the 8th postnatal day by decapitation and TYR, TRP, 5-HT and 5-HIAA were measured as previously described (*Bourgoin et al.*, 1974). Each value (μg/g or ml) is the mean ± SEM of 8 determinations. % corresponds to the percentage of TRP not bound to serum albumin. * $p < 0.05$ when compared to respective control values.

in preparation). Since the neurological maturation occurs earlier in hypotrophic than in normal rats (*Chanez-Bel*, 1972), one can wonder if it may be related to the prolonged lack of binding of tryptophan in serum leading to a higher rate of 5-HT synthesis in the brain of hypotrophic animals. Although no definitive conclusion can be drawn yet, it must be recalled that, in contrast, treatment known to reduce the 5-HT synthesis during the early life period triggers a significant delay in the brain maturation (*Hole*, 1972). In light of the proposed non-transmitter role of 5-HT during brain growth (see *Baker* and *Quay*, 1969), the particular situation in hypotrophic animals should be further explored.

In control as well as in hypotrophic animals, our studies clearly demonstrate that brain tryptophan levels are always correlated with the concentration of free tryptophan in serum. Studies on the ontogenic development of tryptophan transport from blood to brain, therefore, confirm the conclusion drawn from experiments using drugs (*Gessa* and *Tagliamonte*, 1974) and stressful situations (*Knott et al.*, 1973). During development, *i.e.* when the rate of protein synthesis is very high in the brain, the peripheral mechanisms of tryptophan transport are perfectly adapted since the poor binding of the amino acid onto the serum albumin makes it more readily disposable for the cell machinery.

2. Characteristics of the Tryptophan Carrier in Brain Tissues During Development

Previous studies with slices from various areas (brain stem, cerebellum, raphé) have shown that the uptake of tryptophan in brain tissues occurs at a faster rate around the 1—3rd postnatal weeks than in adult rats (*Bourgoin et al.*, 1974; *Bourgoin*, 1976). Similar experiments using crude (P_2 fraction) or purified (B band) synaptosomes from the brain stem confirmed this conclusion: as shown in Fig. 3, the V_{max} of the high affinity carrier for tryptophan (apparent K_m in the range of 10^{-5} M) in P_2 fraction was maximum during the 2nd postnatal week. These developmental changes were not confined to the posterior brain since they are also observed in synaptosomal preparations from the hypothalamus: on the 14th postnatal day, the accumulation of tryptophan (in nmol per mg protein) was 285 % that found with synaptosomes from adult rat (*Bourgoin*, 1976). In contrast to these important modifications of the velocity, the apparent affinity of the membrane carrier for tryptophan exhibited very discrete changes during ontogenesis: in brain stem synaptosomes, the apparent K_m was slightly different from the adult value for only the early postnatal period (Fig. 4). The slight reduction observed (—35 %) might indicate that the affinity of the membrane carrier for tryptophan is better in younger rats. This finding confirms that obtained in preliminary experiments using tissue slices (*Bourgoin et al.*, 1974).

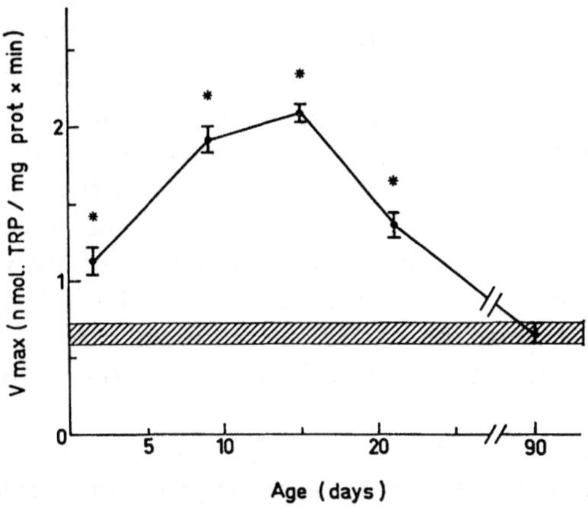

Fig. 3. Changes in the V_{max} of TRP uptake in synaptosomes from the rat brain stem during development. The crude synaptosomal (P_2) fraction was used at all ages. The uptake of TRP was measured at the end of a 30 sec incubation at 37 °C with ^{3}H-TRP (5 μM—40 μM) as previously described (*Bourgoin et al.*, 1977 a). V_{max} (in nmol TRP accumulated per mg synaptosomal protein and per min) was determined by linear regression analysis of double reciprocal plots (1/v versus 1/TRP concentration). Each point corresponds to the mean ± SEM of 3 independent determinations. * $p < 0.05$ when compared to the adult (90 day-old) value

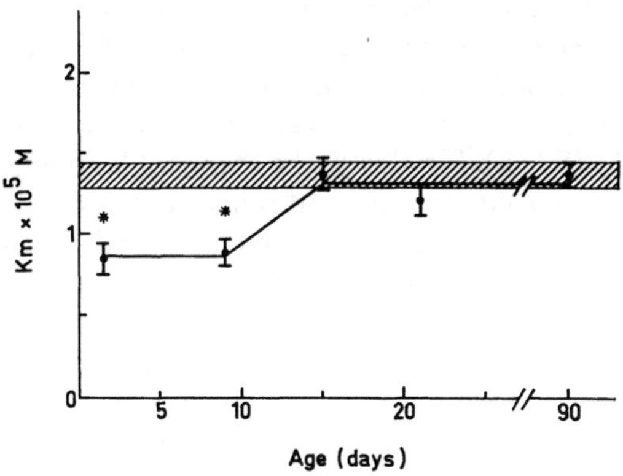

Fig. 4. Ontogenetic changes in the apparent K_m of TRP carrier in synaptosomes from the rat brain stem. The experiments are those described in the legend to Fig. 3. Each point is the mean ± SEM of 3 independent determinations. * $p < 0.05$ when compared to the adult value

According to *Denizeau* and *Sourkes* (1977), tryptophan uptake occurs preferentially in serotoninergic terminals since the electrolytic lesion of the midbrain raphé in adult rats not only produces a 60 % reduction in 5-HT uptake in forebrain synaptosomes, but also a 37 % decrease in the synaptosomal accumulation of tryptophan in the hypothalamus and the hippocampus. Accordingly, this indicates that 37 % of the accumulated tryptophan is normally taken up by 60 % of the serotoninergic terminals, *i.e.* that 0.37/0.60 = 62 % of the whole tryptophan is accumulated selectively in the serotoninergic units of a synaptosomal preparation. Since in the richest zone in the CNS, the suprachiasmatic nucleus, only 5 % of terminals are serotoninergic (*Kuhar* and *Aghajanian*, 1973), the data reported by *Denizeau* and *Sourkes* (1977) lead to the proposal that, in the hippocampus and the hypothalamus, less than 5 % of nerve endings (the serotoninergic ones) accumulate 62 % of the whole tryptophan. Therefore the tryptophan carrier in the membrane of one serotoninergic terminal would be at least 30 times more active than that of any other terminal.

During development, this would result in a close correlation between the increase in the V_{max} of tryptophan uptake in synaptosomes and the proliferation of serotoninergic terminals in a given area. Indeed, such a relation seems to occur since the ontogenetic evolution of tryptophan uptake closely resembles that of the high affinity uptake process for 5-HT in synaptosomal fractions from the brain stem for instance (*Bourgoin*, 1976). However, further experiments completely rule out the idea of a preferential uptake of tryptophan in serotoninergic nerve endings in young animals: as shown in Table 2, the electrolytic raphé lesion in 6 day-old rats induced the specific degeneration of 60—80 % of serotoninergic terminals but failed to alter the kinetic characteristics of tryptophan uptake in synaptosomes from various forebrain areas (see also *Bourgoin et al.*, 1977 a). Similarly, the chemical lesioning of descending serotoninergic neurons by the administration of 5, 7-dihydroxytryptamine (100 mg/kg s.c.) on the 1st and 2nd postnatal days did not result in any significant decrease in the accumulation of tryptophan in synaptosomes from the spinal cord (Table 2). The lack of correlation between tryptophan and 5-HT uptake mechanisms in synaptosomes is further confirmed by examining what occurred in the brain stem of 2 week-old animals which had been treated with 5, 7-dihydroxytryptamine at birth: as shown in Table 2, this treatment induced a significant increase in the V_{max} of the synaptosomal 5-HT uptake process without any change in the kinetic characteristics of the tryptophan carrier. In spite of the increased number of serotoninergic terminals in the brain stem of 5, 7-dihydroxytryptamine-treated rats, the velocity of tryptophan uptake process in synaptosomes from this area was equal to that found in similar preparations from pair-aged controls (Table 2).

These observations are in agreement with previous data obtained with cultured cell lines (*Bauman et al.*, 1974) showing that the high affinity carrier for tryptophan (K_m ranging between 7 μM and 59 μM) is as active in fibroblasts, C_6 glial cells, neuroblastoma clones as it is in crude or purified synaptosomal fractions containing serotoninergic nerve endings.

In conclusion, the high capacity of slices or synaptosomes to accumulate tryptophan does not correspond to the proliferation of serotoninergic terminals in the brain during development. The very high activity of the tryptophan carrier is likely a particularity exhibited by the membranes of most cells in the growing brain. With the lack of binding in plasma, this is a second mechanism perfectly adapted to supplying tissues with enough tryptophan for supporting a high rate of protein synthesis.

Table 2. *Effects of degeneration or sprouting of serotoninergic terminals on the characteristics of synaptosomal TRP uptake in various brain areas in young rats*

			5-HT uptake		TRP uptake	
			K_m	V_{max}	K_m	V_{max}
A	Cortex	Control	58.1	9.20	23.3	1.92
		Raphé lesioned	60.2	2.04 (—78 %)	23.3	1.99
	Hypothalamus	Control	58.8	21.86	22.8	3.42
		Raphé lesioned	63.6	7.95 (—64 %)	24.7	3.41
B	Spinal cord	Control	77.4	19.35	14.8	1.29
		5, 7-HT	81.0	1.19 (—94 %)	14.8	1.33
	Brain stem	Control	70.2	16.77	16.9	2.03
		5, 7-HT	84.3	26.55 (+58 %)	18.2	2.12

A. *Effects of midbrain raphé lesion:* Electrolytic lesion of dorsal and median raphé was performed on the 6th postnatal day and rats were killed when 60 day-old. Uptake was measured on P_2 fractions (*Bourgoin et al.*, 1977 a) with triplicate determinations for each concentration of the labelled substrate (^{3}H-5-HT or ^{3}H-TRP). *5-HT uptake:* K_m in nM and V_{max} in pmoles ^{3}H-5-HT accumulated per mg protein and per min; *TRP uptake:* K_m in μM and V_{max} in nmoles ^{3}H-TRP accumulated per mg protein and per min.

B. *Effects of 5, 7-dihydroxytryptamine treatment:* 5, 7-HT (100 mg/kg s.c.) was administered twice on the first and the second postnatal day. Young rats were killed when 2 week-old. 5-HT and TRP uptake were measured in P_2 fractions as described above (A).

As emphasized by several authors (*Fernstrom* and *Wurtman*, 1972; *Pérez-Cruet et al.*, 1974), the uptake of tryptophan in tissues is also modulated by the competition exerted by several neutral amino acids such as tyrosine, phenylalanine, leucine ... Since the levels of tyrosine in serum double during the first two weeks following birth (*Bourgoin*, 1976), they may also contribute to the progressive decrease in the concentration of tryptophan in the CNS during this period (Fig. 1). However, the comparison of Fig. 1 and 2 and the data in Table 1 indicate that the binding of tryptophan in serum is apparently the main factor controlling the accumulation of the amino acid in the brain of developing rats. This is not surprising since *Curzon* and *Knott* (1977) and *Gessa* and *Tagliamonte* (1974) also came to the same conclusion in adult animals.

3. The Role of Tryptophan in the Regulation of 5-HT Synthesis During Development in the Rat

Tryptophan loading (200 mg/kg i.p.) is well known to induce a pronounced increase in the rate of 5-HT synthesis in the CNS of adult rats (*Bourgoin et al.*, 1974). In contrast, similar treatment only results in a very slight elevation of 5-HIAA levels in the brain of newborn animals indicating that the rate limiting enzyme, tryptophan hydroxylase, is probably saturated by its substrate during the early life period (*Bourgoin et al.*, 1974; *Hoff et al.*, 1976). Experiments with parachlorophenylalanine (300 mg/kg i.p.) confirm this conclusion since the reduction in tryptophan levels (—20 %) occurring on the third hour after its administration is associated with a significant decrease in 5-HT levels only in adult rats; although the effect of parachlorophenylalanine treatment is more pronounced on brain tryptophan levels (—28 %), the latter are still high enough to saturate tryptophan hydroxylase and no reduction in 5-HT levels is detected in neonates (*Bourgoin*, 1976; *Hoff et al.*, 1977). Consequently, in contrast to what occurs in adult rats (see *Hamon* and *Glowinski*, 1974), the synthesis of 5-HT in the CNS of newborn rats during the first postnatal week does not depend on possible fluctuations of tryptophan levels in brain or in serum. The maintenance of a stable and high rate of 5-HT synthesis during the early life period may be relevant to the putative role of this neurotransmitter during the early stages of brain maturation (see *Baker* and *Quay*, 1969).

In conclusion, the two mechanisms limiting the penetration of tryptophan in the brain of adult animals, *i.e.* the binding of the amino acid onto serum albumin in plasma and the carrier mechanism in cell membranes, are perfectly adapted to supplying the brain with large amounts of tryptophan during the early life period. The physiological importance of these adaptations must be emphasized since they support the high synthesis rates of protein and 5-HT (as a growth factor?) occurring in the developing brain.

Acknowledgements

This research was supported by grants from INSERM, DGRST, DRME and la Société des Usines Chimiques Rhône-Poulenc.

References

Baker, P. C., Quay, W. B.: 5-Hydroxytryptamine metabolism in early embryogenesis and the development of brain and retinal tissues. A review. Brain Res. *12*, 273—295 (1969).

Barondes, S. H.: Do tryptophan concentrations limit protein synthesis at specific sites in the brain? In: Aromatic amino acids in the brain, Ciba Found. Symp. (*Wolstenholme, G. E., Fitzsimons, D. W.*, eds.), 22, pp. 265—274. Amsterdam-London-New York: Elsevier, Excerpta Medica, North Holland. 1974.

Bauman, A., Bourgoin, S., Benda, P., Glowinski, J., Hamon, M.: Characteristics of tryptophan accumulation by glial cells. Brain Res. *66*, 253—263 (1974).

Bourgoin, S.: Evolution du métabolisme cérébral de la sérotonine chez le rat au cours du développement. Thèse de Doctorat ès Sciences Naturelles, Université de Paris VII, 1—303 (1976).

Bourgoin, S., Enjalbert, A., Adrien, J., Héry, F., Hamon, M.: Midbrain raphe lesion in the new born rat: II. Biochemical alterations in serotoninergic innervation. Brain Res. *127*, 111—126 (1977 a).

Bourgoin, S., Faivre-Bauman, A., Benda, P., Glowinski, J., Hamon, M.: Plasma tryptophan and 5-HT metabolism in the CNS of the new born rat. J. Neurochem. *23*, 319—327 (1974).

Bourgoin, S., Faivre-Bauman, A., Héry, F., Ternaux, J. P., Hamon, M.: Characteristics of tryptophan binding in the serum of the new born rat. Biol. Neonate *31*, 141—154 (1977 b).

Chanez-Bel, C.: Retard de croissance intrautérine chez le rat. Thèse, Paris VI, 1972.

Curzon, G., Knott, P. J.: Environmental, toxicological and related aspects of tryptophan metabolism with particular reference to the central nervous system. CRC Critical Rev. Toxicol. *1977*, 145—187.

De Montis, M. G., Olianas, M. C., Mulas, G., Tagliamonte, A.: Evidence that only free serum tryptophan exchanges with the brain. Pharmacol. Res. Comm. *9*, 215—220 (1977).

Denizeau, F., Sourkes, T. L.: Regional transport of tryptophan in rat brain. J. Neurochem. *28,* 951—959 (1977).

Etienne, P., Young, S. N., Sourkes, T. L.: Inhibition by albumin of tryptophan uptake by brain. Nature *262,* 144—145 (1976).

Fernstrom, J. D., Wurtman, R. J.: Brain serotonin content: physiological regulation by plasma neutral amino acids. Science *178,* 414—416 (1972).

Gessa, G. L., Tagliamonte, A.: Serum free tryptophan: control of brain concentrations of tryptophan and of synthesis of 5-hydroxytryptamine. In: Aromatic amino acids in the brain, Ciba Found. Symp. (*Wolstenholme, G. E., Fitzsimons, D. W.,* eds.), 22, pp. 207—216. Amsterdam-London-New York: Elsevier, Excerpta Medica, North Holland. 1974.

Hamon, M., Bourgoin, S., Morot-Gaudry, Y., Héry, F., Glowinski, J.: Role of active transport of tryptophan in the control of 5-hydroxytryptamine biosynthesis. In: Advances in Biochemical Psychopharmacology (*Costa, E., Gessa, G. L., Sandler, M.,* eds.), 11, pp. 153—162. New York: Raven Press. 1974.

Hamon, M., Glowinski, J.: Regulation of serotonin synthesis. Life Sci. *15,* 1533—1548 (1974).

Hoff, K. M., Baker, P. C., Buda, R.: Free tryptophan levels in regions of the maturing mouse brain. Brain Res. *73,* 376—379 (1974).

Hoff, K. M., Baker, P. C., Buda, R. E.: Precursor and end product effects upon indoleamine maturation in mouse brain. Biol. Neonate *29,* 360—367 (1976).

Hoff, K. M., Baker, P. C., Buda, R. E.: Effects of parachlorophenylalanine on indoleamines in maturing mouse brain. Gen. Pharmacol. *8,* 213—215 (1977).

Hole, K.: Reduced 5-hydroxyindole synthesis reduces postnatal brain growth in rats. Eur. J. Pharmacol. *18,* 361—366 (1972).

Knott, P. J., Joseph, M. H., Curzon, G.: Effects of food deprivation and immobilization on tryptophan and other amino acids in the rat brain. J. Neurochem. *20,* 249—251 (1973).

Korpi, E. R., Oja, S. S.: Tryptophan influx into rat brain slices in different buffers and in the presence of other amino acids and albumin. J. Neurochem. *30,* 925—927 (1978).

Kuhar, M. J., Aghajanian, G. K.: Selective accumulation of ^{3}H-serotonin by nerve terminals of raphe neurons: an autoradiographic study. Nature (New Biol.) *241,* 187—189 (1973).

Miller, M., Leahy, J. P., Stern, W. C., Morgane, P. J., Resnick, O.: Tryptophan availability: relation to elevated brain serotonin in developmentally protein-malnourished rats. Exp. Neurol. *57,* 142—157 (1977).

Pérez-Cruet, J., Chase, T. N., Murphy, D. L.: Dietary regulation of brain tryptophan metabolism by plasma ratio of free tryptophan and neutral amino acids in humans. Nature *248,* 693—695 (1974).

Sidransky, H., Sarma, D. S. R., Bongiorno, M., Verney, E.: Effect of dietary tryptophan on hepatic polyribosomes and protein synthesis in fasted mice. J. Biol. Chem. *243,* 1123—1132 (1968).

Tyce, G. M., Flock, E. V., Owen, C. A., jr.: Tryptophan metabolism in the brain of the developing rat. In: Progress in Brain Research (*Himwich, W. A., Himwich, H. E.*, eds.), 9, pp. 198—203. Amsterdam: Elsevier. 1964.

Wigglesworth, J. S.: Experimental growth retardation in the fetal rat. J. Path. Bact. *88*, 1 (1964).

Authors' address: Dr. *M. Hamon*, Groupe NB-INSERM U 114, Collège de France, 11, place Marcelin Berthelot, F-75231 Paris Cedex 05, France.

Journal of Neural Transmission, Suppl. 15, 107—114 (1979)

Kinetics of Tryptophan Transport Into the Brain

T. L. Sourkes

Departments of Psychiatry and Biochemistry, Faculty of Medicine, McGill University, Montreal, Canada

With 2 Figures

Summary

Tryptophan is transported into the brain by a complex system, experimental and kinetic analysis of which favours a carrier mechanism functioning in association with diffusion (3-parameter model), as distinct from a system comprising two carriers (4-parameter model). The Lineweaver-Burk method is limited as a means of distinguishing between these alternative models. A statistical (linearization) technique represents an improvement over the graphical method in providing the best estimates of the parameters, as well as a measure of the variance (error mean square) of each fit to the data. However, even with this a distinction between alternative models ultimately depends upon analysis of a variety of kinetic experiments.

1. General Considerations

Substances reaching the cells of the brain do so after traversing two sets of membranes: (a) the endothelial wall of the cerebral capillaries, and (b) cellular membranes. The first set represents the blood-brain barrier, and separates the blood plasma from the extracellular fluid of the brain. *Oldendorf* (1971) has devised a technique for studying movement of compounds across this barrier, by application of a double-labeling technique. The second set separates the extracellular fluid from the intracellular water. Movement across these membranes has been studied by the slice technique as well as by uptake into synaptosomal preparations, or often into the crude P_2 fraction. Methods based upon steady-state conditions have been described by *Baños et al.* (1973) and by *Seta et al.* (1973). These measure net uptake representing the balance between movement across the blood-brain barrier, actual uptake of the transported substance into brain cells, and efflux.

This subject has been reviewed recently (*Lajtha*, 1974; *Lorenzo*, 1974; *Pardridge* and *Oldendorf*, 1977; *Young* and *Sourkes*, 1977).

2. Tryptophan Uptake Into Brain

The present paper is concerned with a special aspect of the transport of tryptophan and other substances, namely, the kinetic analysis of uptake data. Although our views on this subject stem from work on the transport of tryptophan in slices of rat-brain cortex (prepared with a Stadie-Riggs microtome) (*Kiely* and *Sourkes*, 1972) and other regions of the brain (prepared with a McIlwain chopper) (*Denizeau* and *Sourkes*, 1977), the analysis is quite general, applicable to other experimental systems as already described by *Denizeau et al.* (*Denizeau*, *Wyse*, and *Sourkes*, 1977).

Transport of tryptophan into cerebral tissue was first described as an active process by *Joanny et al.* (1968) and *Barbosa et al.* (1970). In my laboratory *Kiely* showed that the rate of transport of labeled tryptophan from the bathing medium into the non-inulin space of brain slices conforms to the hyperbolic equation:

$$v = aS^b \qquad (1)$$

where v is the rate of uptake; S is the initial concentration of tryptophan in the external medium; and a and b are constant. The shape of the curve expressed in this function indicated at once that the mechanism for tryptophan uptake is complex. In fact, when graphical methods of analysis were applied, the multiple nature of the mechanism became apparent.

This point is illustrated in Fig. 1 (*cf.* Fig. 2 in *Kiely* and *Sourkes*, 1972). Curve A represents the hyperbolic fit to the observed points over a wide range of tryptophan concentrations (0.2—10 mM). Curve B represents the Michaelis-Menten equation whose constants can be roughly estimated by the graphical method of *Lineweaver* and *Burk* (1934) from the results at the lower range of concentrations (in this case, for values of S = 0.2—3.0 mM). Curve C is simply the arithmetic difference between the other two curves. It has two important features. (i) Through its intersection of the Y-axis below the origin, it provides a measure of the rate of efflux of tryptophan during the course of these experiments, which used a 15-minute incubation period. The mean rate is then estimated to be 1.0 mmol $\cdot$ 1^{-1}

per 15 min. (ii) The other property of this line is that the numerical value of its slope ($v^{-1}/S^{-1} = S/v$) is 1.0. With inulin as an extracellular marker, the inulin space is 19 μl/100 mg of tissue, the non-inulin space is 63 μl/100 mg under our experimental conditions. We have estimated net transport rates in terms of molar concentration moving into the non-inulin space, considered as the intracellular aqueous compartment, per unit time. Because both axes of the graph made use of molarities, it was evident from the unit slope of curve C that above a certain initial concentration of substrate in the nutrient medium, entry of tryptophan was occurring on a strictly proportional basis, *i.e.* by diffusion. The use of other unitages based upon weight of tissue or protein would also have produced a straight line for curve C, but the diffusional nature of this component of the overall uptake process (curve A) would not have been so readily evident.

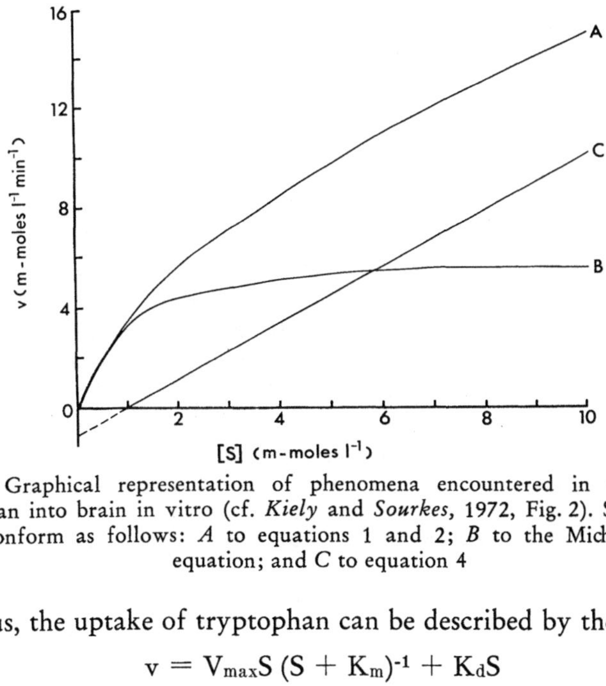

Fig. 1. Graphical representation of phenomena encountered in transport of tryptophan into brain in vitro (cf. *Kiely* and *Sourkes*, 1972, Fig. 2). See text. The curves conform as follows: *A* to equations 1 and 2; *B* to the Michaelis-Menten equation; and *C* to equation 4

Thus, the uptake of tryptophan can be described by the equation:

$$v = V_{max}S\,(S + K_m)^{-1} + K_dS \qquad (2)$$

where v is the velocity of influx; V_{max} is the maximal velocity of saturable transport; K_m is the transport constant, corresponding to the Michaelis-Menten constant; K_d is the diffusion constant; and S is the initial concentration of the amino acid in the medium.

This is the equation developed in the case of tryptophan transport into brain by *Kiely* and *Sourkes* (1972), *Vahvelainen* and *Oja* (1972),

Denizeau and *Sourkes* (1977), and *Denizeau et al.* (1976), and into neuroblastoma clones by *Archer et al.* (1977). A complete mathematical description would have to take into account efflux of substrate during the experimental period. For short periods of incubation, a simple correction factor suffices (*Young* and *Sourkes*, 1977). Application of equation 2 to the data obtained for uptake of tryptophan by cells of the rat hypothalamus, striatum, cerebellum and raphé nuclei region shows that there may be considerable differences in the values of the constants, particularly the diffusion constant (*Denizeau* and *Sourkes*, 1977). It can be shown that at the usual concentrations of tryptophan in the plasma of rat the contribution of carrier transport is considerably greater than that of diffusion, and that the difference is greatest for the raphé nuclei region. This suggests that the chief serotonin-synthesizing cells of the brain are provided with a mechanism for more rigorous control of tryptophan uptake, *i.e.* one in which passive diffusion is at a minimum and transport essentially depends upon a carrier.

3. Kinetic Analysis

In recent years high and low affinity systems have been described in brain for uptake of amines and amino acids thought to act as neurotransmitters. High and low affinity systems have, in fact, been claimed for tryptophan (*Belin, Chouvet*, and *Pujol*, 1974; *Knapp* and *Mandell*, 1972; *Bauman, Bourgoin, Penda, Glowinski*, and *Hamon*, 1974). In most of these reports investigators have formally applied the method of *Lineweaver* and *Burk* (1934); finding a curve that is concave downward for the plot of v^{-1} vs. S^{-1}, they have interpreted this in terms of the model of multiple transport comprising high and low affinity components. This corresponds in enzyme mechanisms to the kinetics of a catalyst with two independent sites for substrate binding:

$$v = V_1S\,(S + K_{m1})^{-1} + V_2S\,(S + K_{m2})^{-1} \qquad (3)$$

The problem in using this graphical method is that equations 2 and 3 in their double-reciprocal forms both show concavity from below, as illustrated in Fig. 2, and the direction to be taken in extrapolating the points obtained for high concentrations of substrate is not necessarily evident by eye-fitting. In fact, the double-reciprocal plot of equation 3 yields an extrapolate that cuts the Y-axis, but for equation 2 the line passes through the origin. Hence, more precise methods of analysis are necessary in making the decision, for two quite different models of transport are involved. In order to avoid an arbitrary choice, *Wyse* and *Denizeau* prepared programmes in Basic that would test the fit of transport data by the least-squares method to the two simplest of multiple transport systems, *viz.*, that comprising two Michaelis-Menten type equations (4-parameter model, equation 3), and that made up of the sum of one Michaelis-Menten equation and a linear diffusion component (3-parameter model, equation 2). The programmed procedures are actually iterative; they start with the graphically estimated parameters and continue until the solution converges *i.e.* until the fractional change in the error mean square and/or the fractional changes in all the parameters in successive iterations reach some specified value.

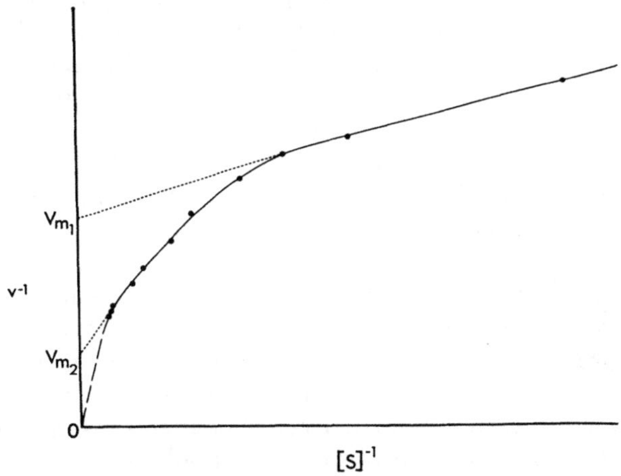

Fig. 2. Graphical representation of the double reciprocal plot of equation 2, showing extrapolation of the curve through the origin, and of equation 3, with two of the possible tangents drawn through the ordinate

A further confirmation of the 3-parameter model for tryptophan uptake in brain has been adduced from the data of *Belin* and *Pujol* (1973) by means of our programmes. This has also been true for the data of *Knapp* and *Mandell* (1972; *Mandell* and *Knapp*, 1977) for uptake of tryptophan into synaptosomes prepared from striatum and septum. They reported both high and low affinity systems. For striatal synaptosomes the K_m for the high affinity transport system was $50 \pm 3\ \mu M$. This mean value corresponds to the values found by *Denizeau* and *Sourkes* (1977) in four regions of rat brain (60 to 300 μM) for the K_m of the active transport component of a system as described by equation 2. The V_{max} estimated by *Knapp* and *Mandell* was 11.1 ± 1 mMol mg^{-1} per 4 min. By converting units, and making reasonable assumptions about the amount of intracellular water making up the weight of synaptosomes, the maximal velocity is equivalent to 0.75—0.90 mMol $l^{-1}min^{-1}$. This is about 5.5—7.0 times the velocity of uptake by blocks of striatal cells (*Denizeau* and *Sourkes*, 1977).

Additional examples of the application of the principle of least-squares fit of data to equations 2 and 3, as well as to a simple diffusional model:

$$v = k_d S \qquad (4)$$

are presented by *Denizeau et al.* (1977). Other investigators have arrived at similar conclusions. Thus, *Archer* and her colleagues (*Archer et al.*, 1977) have found that the tryptophan transport into neuroblastoma cells also conforms best to equation 2. *Atkins* and *Gardner* (1977) have analyzed kinetic data in the literature for goodness-of-fit to these models, as well as to the 2-parameter model (*i.e.* the Michaelis-Menten equation). In no case out of 35 was the 4-parameter model that of "best fit". Furthermore, *Eisenthal* and *Cornish-Bowden* (1974) and *Atkins* and *Nimmo* (1975) have introduced computer programmes to replace the graphical method of *Lineweaver* and *Burk* (1934).

These developments in kinetic measurement should now provide a solid mathematical basis for decision between different models, not only of transport but of enzyme action as well. Thus, there is a formal mathematical resemblance of the 3-parameter model of transport that I have described to a mechanism operating in the case of an enzyme employing two substrates (human liver rhodanese), as described by *Westley* and *Taylor* (1974). In this case the double-reciprocal plot of velocity versus concentration of substrate B yields a curve that is concave downward, the line intercepting the Y-axis closer and closer to the origin as the concentration of substrate A is increased. In the limit, a term that includes A disappears and the curvilinear graph then passes through the origin and is identical in shape to the curve obtained by inversion of equation 2. In fact, the equation adduced by *Westley* and *Taylor* is superimposable on that for the 3-parameter model.

Although the contribution of *Lineweaver* and *Burk* to kinetics is now being updated, it is well to remember that it was these investigators who warned about the caution necessary "in concluding that a reaction follows a certain mechanism because a set of data fits a certain equation . . . Lack of extensive data might also be responsible for a misleading agreement of data with an assigned mechanism." In the case of tryptophan transport as well as other phenomena, a good mathematical fit of a curve to the data does not necessarily define the physiological process. Kinetic analysis can eliminate certain mechanistic possibilities and recommend others although without necessarily distinguishing between them. In that case the ultimate decision between alternate models must be made on the basis of additional experiments (*Denizeau et al.*, 1976).

References

Archer, E. G., Breakfield, X. O., Sharata, M. N.: Transport of tyrosine, phenylalanine, tryptophan and glycine in neuroblastoma clones. J. Neurochem. *28*, 127—135 (1977).

Atkins, G. L., Gardner, M. L. G.: The computation of saturatable and linear components of intestinal and other transport kinetics. Biochim. Biophys. Acta *468*, 127—145 (1977).

Atkins, G. L., Nimmo, I. A.: A comparison of seven methods for fitting the Michaelis-Menten equation. Biochem. J. *149*, 775—777 (1975).

Baños, G., Daniel, P. M., Moorhouse, S. R., Pratt, O. E.: The influx of amino acids into the brain of the rat *in vivo:* the essential compared with some non-essential amino acids. Proc. Roy. Soc. London *183*, 59—70 (1973).

Barbosa, E., Joanny, P., Corriol, J.: Accumulation active du tryptophane dans le cortex cérébral isolé du rat. C.R. Séances Soc. Biol. Paris *164*, 345—350 (1970).

Bauman, A., Bourgoin, S., Benda, P., Glowinski, J., Hamon, M.: Characteristics of tryptophan accumulation by glial cells. Brain Res. *66*, 253 to 263 (1974).

Belin, M. F., Chouvet, G., Pujol, J. F.: Transport synaptosomal du tryptophane et de la tyrosine cérébrale. Stimulation de la vitesse de capture après reserpine ou inhibition de la monoamine oxydase. Biochem. Pharmacol. *23*, 587—596 (1974).

Belin, M. F., Pujol, J. F.: Transport synaptosomal du tryptophane et de la tyrosine cérébrale. Existence de systèmes de capture d'affinité différente. Experientia (Basel) *29*, 411—413 (1973).

Denizeau, F., Sourkes, T. L.: Regional transport of tryptophan in rat brain. J. Neurochem. *28*, 951—959 (1977).

Denizeau, F., Wyse, J., Sourkes, T. L.: Kinetics of multiple transport systems for amino acids in the brain. J. Theor. Biol. *63*, 99—110 (1976).

Eisenthal, R., Cornish-Bowden, A.: The direct linear plot, a new graphical procedure for estimating enzyme kinetic parameters. Biochem. J. *139*, 715—720 (1974).

Joanny, P., Barbosa, E., Corriol, J.: Accumulation active de quelques acides aminés dans les coupes de cerveau du rat. J. Physiol. (Paris), Suppl. *60*, 265 (1968).

Kiely, M., Sourkes, T. L.: Transport of *L*-tryptophan into slices of rat cerebral cortex. J. Neurochem. *19*, 2863—2872 (1972).

Knapp, S., Mandell, A. J.: Narcotic drugs: effects on the serotonin biosynthetic systems of the brain. Science (Washington) *177*, 1209—1211 (1972).

Lajtha, A.: Amino acid transport in the brain *in vivo* and *in vitro*. In: Aromatic Amino Acids in the Brain, Ciba Found. Symp. 22 (new series) (*Wolstenholme, G. E. W., Fitzsimons, D. W.*, eds.), pp. 25—49. Amsterdam: Elsevier. 1974.

Lineweaver, H., Burk, D.: The determination of enzyme dissociation constants. J. Am. Chem. Soc. *56*, 658—666 (1934).

Lorenzo, A. V.: Amino acid transport mechanisms of the cerebrospinal fluid. Fed. Proceedings Fed. Amer. Soc. Exp. Biol. *33*, 2079—2085 (1974).

Mandell, A. J., Knapp, S.: Regulation of serotonin biosynthesis in brain: role of the high affinity uptake of tryptophan into serotonergic neurons. Fed. Proceedings Fed. Amer. Soc. Exp. Biol. *36*, 2142—2148 (1977).

Pardridge, W. M., Oldendorf, W. H.: Transport of metabolic substances through the blood-brain barrier. J. Neurochem. *28*, 5—12 (1977).

Seta, K., Sansur, M., Lajtha, A.: The rate of incorporation of amino acids into brain proteins during infusion in the rat. Biochim. Biophys. Acta *294*, 472—480 (1973).

Vahvelainen, M. L., Oja, S. S.: Kinetics of influx of phenylalanine, tyrosine, tryptophan, histidine and leucine into slices of brain cortex from adult and 7-day-old rats. Brain Res. *40*, 477—488 (1972).

Westley, J., Taylor, H.: Distinction among formal mechanisms that yield double reciprocal plots concave from below. J. Biol. Chem. *249*, 7148 to 7150 (1974).

Young, S. N., Sourkes, T. L.: Tryptophan in the central nervous system: regulation and significance. Adv. Neurochem. *2*, 133—191 (1977).

Author's address: Dr. *T. L. Sourkes*, Department of Psychiatry, McGill University, 1033 Pine Avenue West, Montreal, Canada H3A 1A1.

Journal of Neural Transmission, Suppl. 15, 115—124 (1979)

Structural Specificity of the Carrier for L-Tryptophan in Rat Cerebral Cortex Slices

M. E. Kiely and **T. L. Sourkes**

Department of Psychiatry, McGill University, Montreal, Quebec, Canada

Summary

The effect of a number of commonly occurring amino acids as well as metabolites originating in the pyrrolase pathway of tryptophan catabolism and in the serotonin pathway have been studied with regard to their effect on the net flux of tryptophan into rat brain cerebral cortex slices. Phenylalanine, *p*-chlorophenylalanine, tyrosine, 3, 4-dihydroxyphenylalanine and kynurenine as well as leucine, isoleucine and valine all inhibited the net uptake of tryptophan. It was concluded from these studies that tryptophan is transported into the cell via the large neutral amino acid carrier. Compounds lacking the aliphatic side chain carboxyl or α-amino group have no action on tryptophan transport.

Introduction

The existence of multiple transport systems for amino acid entry into brain slices has been described by *Blasberg* and *Lajtha* (1966) and by *Neame* (1968). These investigators have shown that there are at least six classes of carriers which transport groups of amino acids of similar structure and charge: acidic, small neutral, large neutral, small basic, large basic and GABA (γ-aminobutyric acid). Studies in several laboratories (*Barbosa, Joanny*, and *Corriol*, 1968; *Joanny, Barbosa*, and *Corriol*, 1968; *Barbosa, Joanny*, and *Corriol*, 1970; *Grahame-Smith* and *Parfitt*, 1970; *Kiely* and *Sourkes*, 1972; *Belin* and *Pujol*, 1973; *Denizeau* and *Sourkes*, 1977) have established conclusively that tryptophan is actively accumulated by rat brain slices and synaptosomes and that this accumulation decreases considerably with anaerobiosis, the presence of metabolic inhibitors in the incubation medium or lowering of the incubation temperature. Tryptophan accumulation by cell bodies (slices of grey matter) is sodium-dependent (*Kiely* and *Sourkes*, 1972; *Lajtha* and *Sershen*, 1975) but accumulation by nerve endings (synaptosomes) is inhibited by this cation (*Bruinvels*, 1975). The kinetics of tryptophan transport into brain has been recently reviewed (*Young* and *Sourkes*, 1977).

Several of the above-mentioned studies have described the effect of amino acids on tryptophan uptake. However, to date there has been no definitive report on the specificity of the tryptophan carrier in brain. The acquisition of new data on the effects of some tryptophan analogues and metabolites helps clarify this question, and it is the aim of this paper to describe the structural requirements which are necessary for tryptophan accumulation by rat brain cerebral cortex slices.

Materials and Methods

L-Tryptophan (carboxyl-^{14}C) was obtained from New England Nuclear Corp. (Boston, MA). Non-radioactive *L*-tryptophan was supplied by Matheson, Coleman and Bell (East Rutherford, NJ). *DL*-α-Methyltryptophan monohydrate, kynurenic acid, xanthurenic acid, *DL*-indole-3-lactic acid, nicotinic acid, picolinic acid and indole-3-acetic acid creatinine sulfate were purchased from Sigma Chemical Co. (St. Louis, MO). The following items were bought from Calbiochem (Los Angeles, CA): *L*-3-hydroxykynurenine, anthranilic acid, tryptamine hydrochloride, *L*-5-hydroxytryptophan, 5-hydroxytryptamine creatinine sulfate and 5-hydroxyindoleacetic acid as the cyclohexylammonium salt with methanol. 6-Methyl-*DL*-tryptophan was obtained from Mann Research Laboratories (New York City, NY). Quinaldic acid was a product of Eastern Organic Chemical Distillation Products Industries (Rochester, NY). Nutritional Biochemicals Corp. (Cleveland, OH) provided quinolinic acid. α-Methyl-5-hydroxy-*DL*-tryptophan was given to us by Merck, Sharp and Dohme Research Laboratories (Rahway, NJ) while the Upjohn Company (Kalamazoo, MI) donated α-methyl-5-hydroxytryptamine. All other chemicals were reagent grade and were used without further purification.

Rat brain cerebral cortex slices were prepared and incubated as described elsewhere (*Kiely* and *Sourkes*, 1972). In that study it was established that tryptophan uptake into cerebral cortex slices is linear for the first 15 min of incubation and that the time for tryptophan uptake to attain equilibrium is approximately 60 min. During the incubation period some efflux of tryptophan from the cells takes place, and can be estimated (*Young* and *Sourkes*, 1977) on the data of *Kiely* and *Sourkes* (1972) to be of the order of 60 μmol per liter of cell water per min. However, in the experiments reported here the data are expressed simply as the net accumulation of tryptophan under the described conditions (μmol of tryptophan per ml of intracellular water per 15 min).

Table 1. *Effect of tryptophan metabolites with the pyrrole ring intact on the accumulation of L-tryptophan by incubated slices of rat cerebral cortex*[b]

Initial concentration (mM) Tryptophan	Metabolite added	Initial concentration (mM) metabolite	No. of detns.	Tryptophan accumulation μmoles/ml cell water	p[c]
0.06	—	—	4	0.40 ± 0.02	
0.06	5-hydroxytryptophan	0.02	4	0.39 ± 0.02	> 0.05
0.06	5-hydroxytryptophan	0.05	4	0.36 ± 0.01	> 0.05
0.2	—	—	4	1.07 ± 0.10	
0.2	5-hydroxytryptophan	0.05	4	1.09 ± 0.09	> 0.05
0.2	5-hydroxytryptophan	0.5	4	0.78 ± 0.03	< 0.05
1.0	—	—	3	3.10 ± 0.10	
1.0	5-hydroxytryptophan[a]	1.0	4	3.49 ± 0.15	> 0.05
1.0	—	—	6	3.13 ± 0.31	
1.0	5-hydroxytryptamine (creatinine sulfate complex)	1.0	8	3.16 ± 0.15	> 0.05
1.0	creatinine sulfate	1.0	4	3.19 ± 0.18	> 0.05
1.0	—	—	3	3.10 ± 0.10	
1.0	5-hydroxyindoleacetic acid	1.0	3	3.27 ± 0.22	> 0.05
1.0	—	—	8	3.28 ± 0.23	
1.0	tryptamine	1.0	8	2.84 ± 0.15	> 0.05
1.0	—	—	4	3.26 ± 0.18	
1.0	indole-3-acetic acid	1.0	4	3.16 ± 0.53	> 0.05
1.0	—	—	4	3.18 ± 0.47	
1.0	*DL*-indole-3-lactic acid	1.0	4	3.11 ± 0.16	> 0.05

[a] Data of *Kiely* and *Sourkes* (1972).

[b] All incubations were 15 min in duration and were conducted at 37 °C in an atmosphere of 100 % oxygen. Indole-3-acetic acid was dissolved in Krebs-Ringer phosphate medium to which ethanol had been added to a final concentration of 3 % (v/v). Unless otherwise noted, only *L*-isomers of the compounds were used. Tabulated data represent mean values ± S.E.

[c] Probability of difference from corresponding control.

Table 2. *Effect of metabolites of the pyrrolase pathway of tryptophan degradation on the accumulation of L-tryptophan by incubated slices of rat cerebral cortex*[b]

Initial concentration (mM) Tryptophan	Metabolite added	Initial concentration (mM) metabolite	No. of detns.	Tryptophan accumulation μmoles/ml cell water	p[c]
1.0	—	—	4	3.26 ± 0.21	
1.0	N-formylkynurenine[a]	1.0	4	2.98 ± 0.25	> 0.05
0.2	—	—	4	0.99 ± 0.03	
0.2	kynurenine	0.05	4	1.01 ± 0.14	> 0.05
0.2	kynurenine	0.2	4	1.06 ± 0.08	> 0.05
0.2	kynurenine	0.5	4	0.86 ± 0.13	> 0.05
0.2	kynurenine	1.0	4	0.94 ± 0.05	> 0.05
0.5	—	—	4	2.24 ± 0.18	
0.5	kynurenine	0.05	4	2.04 ± 0.23	> 0.05
0.5	kynurenine	0.2	4	2.01 ± 0.18	> 0.05
0.5	kynurenine	0.5	4	1.94 ± 0.13	> 0.05
0.5	kynurenine	1.0	4	1.70 ± 0.09	> 0.05
1.0	—	—	4	3.26 ± 0.21	
1.0	kynurenine	0.05	4	2.56 ± 0.26	> 0.05
1.0	kynurenine	0.2	4	2.66 ± 0.19	> 0.05
1.0	kynurenine	0.5	4	2.16 ± 0.20	< 0.01
1.0	kynurenine[a]	1.0	8	2.03 ± 0.15	< 0.001
0.06	—	—	4	0.43 ± 0.06	
0.06	3-hydroxykynurenine	0.05	4	0.34 ± 0.02	> 0.05

0.5	—	—	4	2.24 ± 0.18	
0.5	3-hydroxykynurenine	0.5	4	2.00 ± 0.23	> 0.05
1.0	—	—	4	3.26 ± 0.21	
1.0	3-hydroxykynurenine[a]	1.0	4	2.69 ± 0.33	> 0.05
0.06	—	—	4	0.43 ± 0.06	
0.06	anthranilic acid	0.05	4	0.39 ± 0.02	> 0.05
0.2	—	—	4	0.99 ± 0.03	
0.2	anthranilic acid	0.2	4	1.14 ± 0.15	> 0.05
0.5	—	—	4	2.24 ± 0.18	
0.5	anthranilic acid	0.5	4	2.58 ± 0.13	> 0.05
1.0	—	—	4	3.26 ± 0.21	
1.0	anthranilic acid	1.0	4	3.01 ± 0.18	> 0.05
1.0	—	—	4	3.39 ± 0.16	
1.0	kynurenic acid	0.5	4	3.02 ± 0.15	> 0.05
1.0	xanthurenic acid	1.0	4	3.48 ± 0.30	> 0.05
1.0	—	—	4	3.18 ± 0.47	
1.0	quinolinic acid	1.0	4	3.14 ± 0.08	> 0.05
1.0	quinaldic acid	1.0	3	2.43 ± 0.43	> 0.05
1.0	picolinic acid	1.0	4	3.08 ± 0.17	> 0.05
1.0	nicotinic acid	1.0	4	3.03 ± 0.17	> 0.05

[a] Data of *Kiely* and *Sourkes* (1972).

[b] All incubations were 15 min in duration and were conducted at 37 °C in an atmosphere of 100 % oxygen. Unless otherwise noted, only *L*-isomers of the compounds were used. Tabulated data represent mean values ± S.E.

[c] Probability of difference from corresponding control.

Table 3. *Effect of some tryptophan analogs on the accumulation of L-tryptophan by incubated slices of rat cerebral cortex*[b]

Initial concentration (mM) Tryptophan	Analog added	Initial concentration (mM) analog	No. of detns.	Tryptophan accumulation μmoles/ml cell water	p[c]
0.06	—	—	4	0.40 ± 0.02	
0.06	α-methyl-*DL*-tryptophan	0.04	4	0.33 ± 0.04	> 0.05
0.2	—	—	4	1.07 ± 0.10	
0.2	α-methyl-*DL*-tryptophan	0.10	4	1.09 ± 0.03	> 0.05
0.2	α-methyl-*DL*-tryptophan	1.0	4	0.88 ± 0.11	> 0.05
1.0	—	—	3	3.10 ± 0.10	
1.0	α-methyl-*DL*-tryptophan	1.0	4	3.73 ± 0.28	> 0.05
1.0	—	—	4	2.89 ± 0.20	
1.0	α-methyl-5-hydroxytryptamine	1.0	4	3.77 ± 0.06	< 0.01
1.0	—	—	4	3.39 ± 0.16	
1.0	4-methyl-*DL*-tryptophan[a]	1.0	4	2.97 ± 0.17	> 0.05
1.0	6-methyl-*DL*-tryptophan	1.0	4	3.19 ± 0.16	> 0.05
1.0	—	—	4	2.89 ± 0.20	
1.0	4-hydroxy-*DL*-tryptophan[a]	1.0	4	3.61 ± 0.24	> 0.05
1.0	—	—	4	3.39 ± 0.16	
1.0	c-methyl-5-hydroxy-*DL*-tryptophan	1.0	4	3.49 ± 0.21	> 0.05

[a] Data of *Kiely* and *Sourkes* (1972).

[b] All incubations were 15 min in duration and were conducted at 37 °C in an atmosphere of 100 % oxygen. Tabulated data represent mean values ± S.E.

[c] Probability of difference from corresponding control.

Table 4. *Influence of amino acids, amines and acids on tryptophan transport into slices of rat cerebral cortex*

Type of compound	Inhibitory	No effect	
Amino acids	phenylalanine	glycine	alanine
	p-chlorophenylalanine[a]	threonine	serine
	tyrosine	histidine	proline
	3, 4-dihydroxyphenylalanine	hydroxyproline	cysteine
	leucine	glutamic acid	γ-aminobutyric acid
	isoleucine	glutamine	arginine
	valine	lysine	3-hydroxykynurenine
	5-hydroxytryptophan	formylkynurenine	
	kynurenine	α-methyltryptophan[a]	
		4-methyltryptophan[a]	
		6-methyltryptophan[a]	
		4-hydroxytryptophan[a]	
		α-methyl-5-hydroxytryptophan[a]	
		methionine[b]	
Amines		tryptamine	
		5-hydroxytryptamine	
Acids		indoleacetic acid	
		indolelactic acid	
		5-hydroxyindoleacetic acid	
		anthranilic acid	
		quinaldic acid	
		kynurenic acid	
		xanthurenic acid	
		quinolinic acid	
		picolinic acid	
		nicotinic acid	

[a] Racemic compound was used.
[b] See reference to *Grahame-Smith* and *Parfitt* (1970).

Results

The effects of a number of tryptophan metabolites on tryptophan accumulation by cerebral cortex slices are presented in the data of Tables 1 and 2. The metabolites have been grouped in these two tables according to their structural similarities. The results in Table 1 show that of all the indole metabolites studied only 5-hydroxytryptophan was inhibitory, although not at all concentrations of tryptophan and 5-hydroxytryptophan used. The other indole-containing amines and acids did not affect the uptake of tryptophan in equimolar concentration. Metabolites of the pyrrolase pathway of tryptophan degradation (Table 2) also did not exert a significant effect on tryptophan accumulation by cerebral cortex slices, with the exception of kynurenine. Concentrations of this amino acid at or near the concentration of tryptophan in the medium significantly decreased accumulation of that amino acid.

The effects of various tryptophan analogs on the accumulation of tryptophan are shown in Table 3. One compound appeared to stimulate tryptophan uptake: α-methyl-5-hydroxytryptamine. The remaining analogs α-methyltryptophan, α-methyl-5-hydroxytryptophan, 4-hydroxytryptophan, 4-methyltryptophan and 6-methyltryptophan had no effect on tryptophan accumulation.

Discussion

In a previous paper (*Kiely* and *Sourkes*, 1972) and in the three tables above we have attempted to detect those structural features which favor inhibition of tryptophan transport into cerebral cortex slices of the rat. A summary of this material is presented in Table 4. The only compounds which inhibited tryptophan accumulation were phenylalanine, *p*-chlorophenylalanine, tyrosine, 3, 4-dihydroxyphenylalanine, 5-hydroxytryptophan, leucine, isoleucine, valine and kynurenine. These represent the aromatic amino acids and the large neutral amino acids. Some investigators have found that methionine fits into the latter group (*Grahame-Smith* and *Parfitt*, 1970), although methionine had no inhibitory action in our previous work. The amines and acidic compounds tested (Table 4) had no effect. However, α-methyl-5-hydroxytryptamine had a stimulatory action on tryptophan uptake, when both compounds were used in equimolar concentrations. This effect cannot be explained at present.

It is interesting that the α-hydrogen of 5-hydroxytryptophan is important to the inhibitory action of that compound, for its substitution by a methyl group resulted in the disappearance of inhibitory action under the conditions tested. α-Methyltryptophan was also not inhibitory.

These results indicate that tryptophan transport into brain, as studied by the slice technique *in vitro* is affected significantly only by the aromatic amino acids and the large neutral amino acids (*Blasberg* and *Lajtha*, 1965; *Blasberg* and *Lajtha*, 1966; *Green* and *Curzon*, 1970; *Barbosa, Herreros,* and *Ojeda*, 1971; *Kiely* and *Sourkes*, 1972; *Joanny, Natali, Hillman,* and *Corriol*, 1973). This is the same group that inhibits the movement of tryptophan across the blood-brain barrier, *i.e.* from blood to the interstitial fluid of the brain (*Pardridge*, 1977). In the case of the blood-brain barrier, α-methyltyrosine also served as a competing amino acid.

Kynurenine offered some competition to tryptophan in the studies of this laboratory. One would have expected that N′-formylkynurenine, and 3-hydroxykynurenine as well, would interact with the tryptophan carrier. For N′-formylkynurenine it may be that the substitutent on the ortho-amino group is large enough to hinder sterically interaction of this compound with the carrier. Similarly, 3-hydroxykynurenine with *o*-amino and *m*-hydroxyl groups may be too large structurally to interact with the carrier.

Acknowledgement

This work was supported by a grant to T.L.S. from the Medical Research Council of Canada.

References

Barbosa, E., Joanny, P., Corriol, J.: Uptake of some amino acids by rat brain slices: effect of various substrates. Experientia *24*, 1196—1197 (1968).

Barbosa, E., Joanny, P., Corriol, J.: Accumulation active du tryptophane dans le cortex cérébral isolé du rat. Comptes rendus des séances de la Société de Biologie *164*, 345—350 (1970).

Barbosa, E., Herreros, B., Ojeda, J. L.: Amino acid accumulation by brain slices: interactions among tryptophan, phenylalanine and histidine. Experientia *27*, 1281—1282 (1971).

Belin, M.-F., Pujol, J.-F.: Transport synaptosomal du tryptophane et de la tyrosine cérébrale. Existence de systèmes de capture d'affinité différente. Experientia *29*, 411—413 (1973).

Blasberg, R., Lajtha, A.: Substrate specificity of steady-state amino acid transport in mouse brain slices. Archiv. Biochem. Biophys. *112*, 361—377 (1965).

Blasberg, R., Lajtha, A.: Heterogeneity of the mediated transport systems of amino acid uptake in brain. Brain Res. *1,* 86—104 (1966).
Bruinvels, J.: Role of sodium in neuronal uptake of monoamines and amino acid precursors. Nature *257,* 606—607 (1975).
Denizeau, F., Sourkes, T. L.: Regional transport of tryptophan in rat brain. J. Neurochem. *28,* 951—959 (1977).
Grahame-Smith, D. G., Parfitt, A. G.: Tryptophan transport across the synaptosomal membrane. J. Neurochem. *17,* 1339—1353 (1970).
Green, A. R., Curzon, G.: The effect of tryptophan metabolites on brain 5-hydroxytryptamine metabolism. Biochem. Pharmacol. *19,* 2061—2068 (1970).
Joanny, P., Barbosa, E., Corriol, J.: Accumulation active de quelques acides aminés dans les coupes de cerveau de rat. Journal de Physiologie (Paris) Suppl. *60,* 265 (1968).
Joanny, P., Natali, J.-P., Hillman, H., Corriol, J.: The uptake and efflux of L-phenylalanine and L-tyrosine from rat brain cerebral cortex slices. Biochem. J. *136,* 77—82 (1973).
Kiely, M., Sourkes, T. L.: Transport of *L*-tryptophan into slices of rat cerebral cortex. J. Neurochem. *19,* 2863—2872 (1972).
Lajtha, A., Sershen, H.: Inhibition of amino acid uptake by the absence of Na^+ in slices of brain. J. Neurochem. *24,* 667—672 (1975).
Neame, K. D.: Progress in Brain Research. In: Brain Barrier Systems. A comparison of the transport systems for amino acids in brain, intestine, kidney and tumor, Vol. 29, p. 185. Amsterdam: Elsevier. 1968.
Pardridge, W. M.: Kinetics of competitive inhibition of neutral amino acid transport across the blood-brain barrier. J. Neurochem. *28,* 103—108 (1977).
Young, S. N., Sourkes, T. L.: Tryptophan in the central nervous system: regulation and significance. Adv. Neurochem. *2,* 133—191 (1977).

Authors' address: Dr. *M. E. Kiely,* Department of Psychiatry, The Montreal General Hospital, 1650 Cedar Avenue, Montreal, Quebec, Canada H3G 1A4.

Journal of Neural Transmission, Suppl. 15, 125—137 (1979)

Interactions Between Tryptophan, Phenylalanine and Sugar Transport in the Small Intestinal Mucosa

J. W. L. Robinson and **F. Alvarado**

Département de Chirurgie Expérimentale, Centre Hospitalier Universitaire Vaudois, Lausanne, Switzerland, and Centre de Recherches sur la Nutrition (C.N.R.S.), Meudon-Bellevue, France

With 4 Figures

Summary

The kinetics of the influx of tryptophan and phenylalanine into guinea-pig intestinal rings have been examined. The transfer of these two amino acids can be described by a single transport system, each amino acid having an affinity constant, K_t, of about 4 mM for the influx mechanism. Mutual inhibition studies have shown that the inhibitory constant of each of the amino acids is also 4 mM.

Although fully competitive inhibition between the two amino acids occurs, the inhibition of the influx of the amino acids by sugars exhibits kinetics of the "pseudo-competitive" type. Such behaviour is compatible with an allosteric interaction between two different binding sites, one for each class of compounds. The lack of correlation between the inhibitory potency of a given sugar and its rate of transfer, as testified by a comparison of the effects of galactose and β-methyl-glucoside on phenylalanine influx, can be reconciled with the "allosteric-interaction hypothesis", but specifically repudiates any theory that attempts to explain such interactions in a way that requires such a correlation. The fact that allosteric interactions are retained in cells preloaded with sodium also precludes a primary role for sodium in the mechanism of such interactions.

Introduction

It is generally believed that all neutral amino acids are absorbed in the intestinal mucosa by a single, sodium-dependent transport process. This assumption has been implicit in a number of kinetic studies concerned with the molecular mechanisms involved in the influx of different amino acids into the enterocyte (*Curran et al.*, 1967; *Alvarado* and *Mahmood*, 1974; *Robinson* and *Sepúlveda*, 1977), but has not been explicitly verified in the tissues examined. Recently, compelling evidence has been obtained from inhibitory studies in favour of the multiplicity of transport systems for neutral amino acids in the rabbit ileum (*Sepúlveda* and *Smith*, 1978), one of the preparations employed in the above-mentioned kinetic studies (*Curran et al.*, 1967). It therefore became imperative to re-examine the properties of neutral amino acid transport in the guinea-pig intestine, the tissue employed for the other kinetic studies cited (*Alvarado* and *Mahmood*, 1974; *Robinson* and *Sepúlveda*, 1977), particularly since the kinetic model derived in these investigations differed from that developed for the rabbit ileum. In the first section of the present communication, evidence is provided in favour of a single transport site in guinea-pig ileum that is shared by tryptophan and phenylalanine.

Monosaccharides are transported into the enterocyte by a separate but parallel sodium-dependent system. It has been known for many years that interactions occur between sugars and amino acids for transport across the intestinal brush border (*Newey* and *Smyth*, 1964; *Alvarado*, 1966; *Chez et al.*, 1966; *Semenza*, 1971), but the reasons for such interactions have been hotly disputed. The arguments have been reviewed in detail in a recent publication (*Robinson* and *Alvarado*, 1977) and need not be repeated here; suffice it to say that the controversy has been reduced to resolving between two possible mechanisms. In the first hypothesis (*Alvarado*, 1966), an allosteric interaction between the two groups of substrates is proposed, due to the proximity of their respective binding sites in the membrane matrix. According to the second hypothesis, the inhibition of net influx of the substrate is believed to occur as a result of an acceleration of its efflux from the cell; the simultaneous influx of the heterologous inhibitor is supposed to create local changes in the intracellular sodium concentration which would favour the return of the substrate to the external medium (*Chez et al.*, 1966; *Semenza*, 1971; *Murer et al.*, 1975). In the second section of this paper, new experiments will be described which tend to distinguish between the merits of the two rival hypotheses.

Methods

The experiments were performed with rings of guinea-pig ileum using conventional techniques (*Alvarado* and *Mahmood*, 1974). The tissues were incubated for two minutes in a ^{14}C-labelled solution of the substrate in Krebs bicarbonate buffer; when concentrations of amino acid greater than 20 mM were employed, the solutions were balanced by adding equivalent quantities of a solution of mannitol to maintain constant the osmolarity and sodium concentration of all media. After the incubation, the tissues were rinsed, blotted, weighed and dissolved separately in 30 % KOH for counting in a liquid scintillation spectrometer (*Robinson* and *Felber,* 1965). The uptake was calculated by comparison with the specific activity of the incubation medium determined under similar conditions.

Kinetic analysis of the influx of an amino acid was performed by fitting the results to the equation:

$$J = \frac{J_{max} \cdot [S]}{K_t + [S]} + K_D \cdot [S] \qquad (1)$$

by systematic iteration to minimise the χ^2 residues between the theoretical and observed values at each substrate concentration, where χ^2 is defined as:

$$\chi^2 = \sum \left(\frac{J_o - J_c}{\sigma}\right)^2. \qquad (2)$$

J_o being the mean observed influx at a given substrate concentration, σ its standard error and J_c the calculated value according to Equation 1. This minimisation procedure also permits the generation of a variance-covariance matrix which provides estimates of the reliability of the values of the constants that emerge (*Cleland,* 1967).

A stringent way of testing whether an amino acid is transported by a single site is provided by an examination of the inhibition of its own uptake (*Alvarado,* 1971; *Sepúlveda* and *Smith,* 1978). If the lowest concentration is considered as substrate and all higher concentrations as "inhibitor", the following relationship (*Inui* and *Christensen,* 1966) should hold for transport via a single site:

$$\frac{1}{1 - \frac{J_i}{J_o}} = \frac{K_i}{K_t}(K_t + [S]) \cdot \frac{1}{[I]} + 1 \qquad (3)$$

or in the special case when inhibitor and substrate are identical:

$$\frac{1}{1 - \frac{J_i}{J_o}} = 1 + (K_t + [S]) \cdot \frac{1}{[I]} \qquad (4)$$

where J_o is the control uptake and J_i is the uptake in the presence of a concentration, [I], of inhibitor.

To examine the competitive inhibition for influx between two amino acids, the effects of different concentrations of the inhibitory amino acid on the influx of a constant concentration of the substrate were analysed with the aid of the following equation (*Neame* and *Richards,* 1972), which is a simple rearrangement of Equation 3:

$$\frac{J_o}{J_i} = 1 + \frac{[I]}{K_i\left(1 + \frac{[S]}{K_t}\right)}. \qquad (5)$$

This is a derivation of the more usual Dixon plot relating $\frac{1}{J}$ to [I], with the advantage that the constant, K_i, can be evaluated directly from the slope of the straight line relating J_o/J_i to [I].

Results and Discussion

The kinetics of the influx of tryptophan into guinea-pig small intestinal rings over a 100-fold concentration range are illustrated in Fig. 1 (upper panel) and the values of the constants derived from the iteration procedure are given in the legend. An estimate of the extracellular space of the preparation was obtained in independent experiments in which intestinal rings were incubated in ^{14}C-labelled mannitol; the uptake of this substrate was a linear function of its concentration, and the slope of the line relating entry to concentration was 0.00372 μmol/100 mg · 2 min · mM. This value agrees reasonably with the K_D for the diffusive component of tryptophan influx (0.0049); it can therefore be tentatively concluded that the process described by K_D involves simply permeation of the extracellular space.

The reasonable fit of the influx kinetics to Equation 1 does not a priori prove that tryptophan is taken up by a single saturable process. However, if the diffusive component is subtracted from all values, and the remaining component of uptake is analysed, then the same values of K_t and J_{max} are obtained in the concentration range 0.5—10 mM as in the concentration range 7—50 mM. When the value of $K_D = 0.00372$ is employed in this analysis, the K_t (evaluated from a weighted regression of a double reciprocal plot) in the lower range of concentrations was 4.8 ± 0.46 mM and in the upper range of concentrations was 7.1 ± 1.94 mM. This agreement, even when applying a K_D derived by independent means, is strongly suggestive of a single transport site for this amino acid. Further support for this conclusion is obtained by plotting these results in accordance with Equation 4. Labelled tryptophan at a concentration of 0.5 mM is considered as substrate and all other concentrations of tryptophan are treated as inhibitors. The results, provided in the lower panel of Fig. 1, obey the expected relationship when a diffusive component equivalent to 0.00372 μmol/100 mg · 2 min · mM is subtracted from each value; the intercept on the abscissa does not differ significantly from 1. The slope of the line can be used to derive a value of K_t which in this case was found to be 4.28 mM.

A similar analysis was applied to the influx of phenylalanine in the same preparation; the details will be published elsewhere. Suffice it to say for the present time that the influx of this amino acid also appears

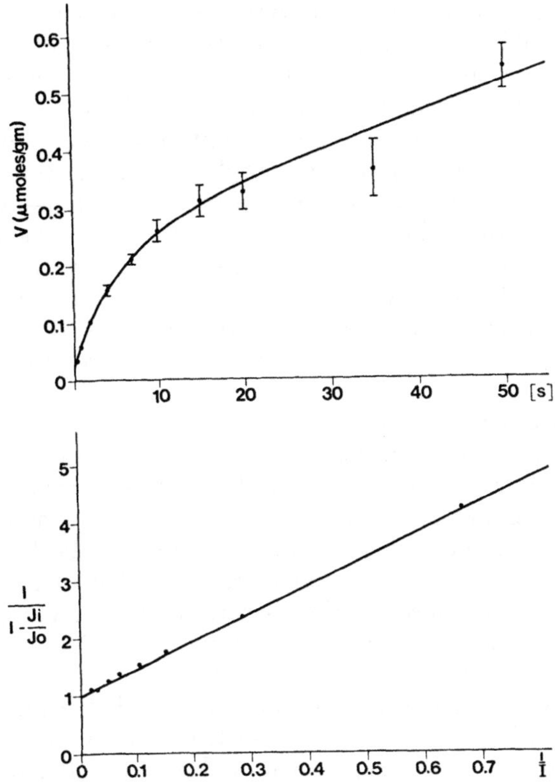

Fig. 1. Kinetics of the influx of L-tryptophan into guinea-pig intestinal rings (upper panel). Tissues were incubated for 2 min in a solution of the required concentration of ^{14}C-labelled tryptophan in Krebs bicarbonate buffer; requisite quantities of mannitol were added to balance the tryptophan concentration, and maintain the osmolarity and sodium concentration constant. The results are the means ± SEM of 8 experiments, and were fitted to Equation 1 by systematic iteration. The best fit was obtained with: $K_t = 4.47 \pm 0.586$ mM; $J_{max} = 0.294 \pm 0.0349$ µmol/100 mg · 2 min; $K_D = 0.0049 \pm 0.00144$ ($\bar{\chi}^2 = 4.221$). These results are replotted (lower panel) in accordance with Equation 4 by first subtracting the diffusive component of entry as determined by independent experiments with mannitol ($K_D = 0.00372$), and then treating 0.5 mM tryptophan as substrate and all other concentrations as inhibitors. The point at [Try] = 1 is omitted as it would give too much weight to the regression line. From the slope, a K_t of 4.28 mM can be calculated

to occur through a single transport site. The parameters obtained by fitting the results to Equation 1 were: $K_t = 4.40 \pm 0.564$, $J_{max} = 0.380 \pm 0.0395$,

$$K_D = 0.00436 \pm 0.00115.$$

In order to examine whether the two amino acids share a common transport pathway, mutual inhibition studies are required. If a fully competitive inhibition occurs between two amino acids for the same site, Equation 5 should hold and the value of K_i obtained for the inhibitor should be the same as the K_t for its own influx. When phenylalanine was substrate and tryptophan was inhibitor, the relationship between J_o/J_i and tryptophan concentration was linear, and the K_i for tryptophan was found to be 4.26 mM—which agrees very well with the value of 4.47 mM for the K_t of the same amino acid. When the reverse experiment was performed, a straight line was again obtained between J_o/J_i for tryptophan and the phenylalanine concentration, and a K_i for phenylalanine of 4.31 mM could be evaluated. The upper panel of Fig. 2 shows these results plotted on a conventional Dixon plot. The straight-line relationship is indicative of fully competitive inhibition.

It thus seems highly probable that tryptophan and phenylalanine share the same transport site in the brush-border membrane of the guinea-pig enterocyte, and that this is the only site available to them for mediated transport. The affinity constants for both amino acids are 4.0—4.5 mM. This conclusion agrees with that of the only other group that has performed a serious kinetic study of tryptophan transport in the intestine: *Munck* and *Rasmussen* (1975) examined the transport of this amino acid in the rat jejunum, also using a wide range of substrate concentrations, and concluded that the amino acid was transported by a single site with an affinity constant of 8 mM.

The inhibition of tryptophan influx by galactose exhibits fundamentally different kinetic properties, in that the relationship between 1/J and [I] is not linear (lower panel of Fig. 2). This behaviour has been designated "pseudo-competitive inhibition" (*Alvarado*, 1967), since examination of the inhibition at variable substrate concentration but constant inhibitor concentration gives rise to curves that are formally identical with competitive inhibition. An experiment of this type concerning the inhibition of phenylalanine influx by a constant concentration of galactose is illustrated in Fig. 3.

This kinetic behaviour, coupled with the fact that the interactions between sugars and amino acids were instantaneous (*Alvarado*, 1968), was instrumental in the development of the "allosteric interaction" theory to explain these interactions (*Alvarado*, 1971). Equations have been developed (*Alvarado,* 1967; *Alvarado* and *Mahmood,* 1974) to describe the modification of the kinetic characteristics of the influx of one substrate in the presence of an allosteric effector, and the predictions of these equations are borne out by the experimental observations (*Robinson* and *Alvarado,* 1977).

This explanation for heterologous interactions between sugars and amino acids, based as it is on kinetic evidence and also on the existence of heterologous countertransport effects (*Robinson* and *Alvarado,* 1971), has not received universal acceptance. Other investigators have preferred to maintain that the unidirectional influx of the substrate is not reduced by the presence of a heterologous inhibitor, but rather that the unidirectional efflux across the brush-border membrane

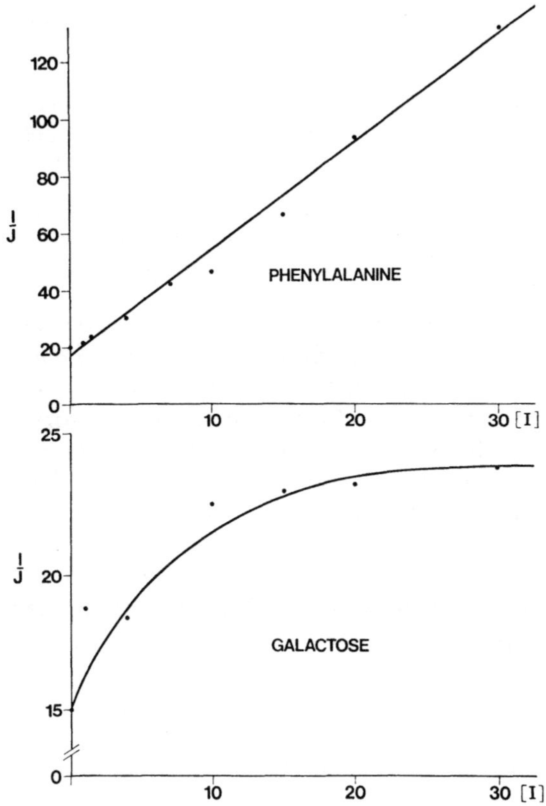

Fig. 2. Inhibition of the influx of L-tryptophan into guinea-pig intestinal rings by L-phenylalanine (upper curve) or D-galactose (lower curve). Results, which are the means of 6 and 8 experiments respectively, are presented on a Dixon plot relating the reciprocal of the influx of tryptophan to the concentration of the inhibitor. A diffusive component of influx with $K_D = 0.0049\ \mu mol/100\ mg \cdot 2\ min \cdot mM$ was subtracted from each value

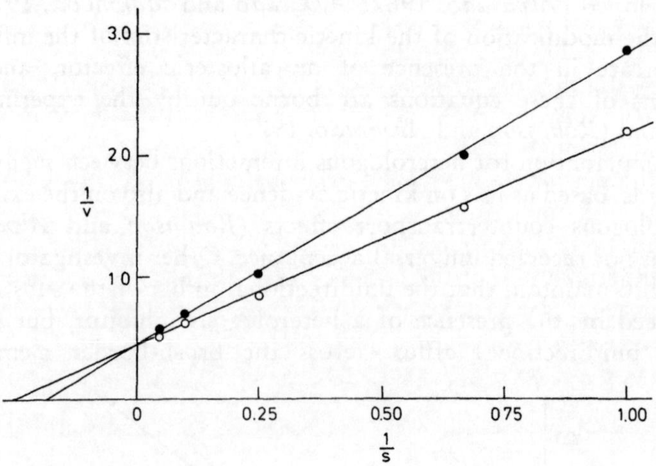

Fig. 3. Kinetics of the inhibition of L-phenylalanine influx by D-galactose. Results, which are the means of six experiments, are corrected for a diffusive component of entry with $K_D = 0.00436$ µmol/100 mg · 2 min · mM and represented on a double reciprocal plot. Weighted regression analysis (*Cleland*, 1967) provides the following values for the kinetic constants: Control series (open circles): $K_t = 4.0 \pm 0.47$; $J_{max} = 0.38 \pm 0.024$: In the presence of 20 mM galactose (closed circles): $K_t = 5.6 \pm 0.38$ mM; $J_{max} = 0.37 \pm 0.007$ µmol/100 mg · 2 min

following its entry into the cell is accelerated (*Chez et at.*, 1966). It has been proposed that this effect is caused by ephemeral changes in the sodium concentration in the immediate vicinity of the carrier at the internal face of the membrane. An alternative formulation of the same basic hypothesis is that a change in the intracellular sodium concentration, provoked by the co-transport of the inhibitor, is responsible for a fall in the rate of influx of the substrate (*Semenza*, 1971). This last possibility can be formally rejected since the unidirectional influx of organic solutes is independent of the intracellular sodium concentration (*Robinson*, 1978). Concerning the first formulation of this hypothesis, we have devised two experimental tests of its validity.

If ephemeral changes in the intracellular sodium-ion concentration were instrumental in causing an acceleration of the return of substrate to the external bathing medium, then heterologous interactions could only occur when the sodium gradient across the brush-border membrane was maintained intact. This postulate was first tested by *Frizzell* and *Schultz* (1971) who killed the intestine by incubating it in cyanide and ouabain, and then proceeded to demonstrate that no heterologous interactions occurred. Their negative evidence was not convincing in view of the harsh treatment meted out to their tissue;

in the present study, we have abolished the sodium gradient across the brush-border membrane by more humane treatment, namely preincubation in the cold under conditions that load the enterocyte with sodium and deplete it of potassium (*Robinson*, 1978), followed by rewarming in the absence of potassium to maintain the reversed gradients, and have shown that heterologous interactions still occur at approximately the same level as seen under control conditions (Table 1). This result is extremely difficult to reconcile with the "accelerated-efflux hypothesis".

A second postulate of the same hypothesis is amenable to experimental test. Clearly, if the interaction between heterologous substrates is secondary to local effects of sodium ions at the internal face of the membrane, sugars inducing the greatest influx of sodium should also be

Table 1. *Effect of a preincubation in melting ice on the interactions between sugars and amino acids for influx into guinea-pig intestinal rings during a subsequent incubation at 37 °C*

	Uptake of substrate (μmol/g fresh tissue)	
	Experiment 1	Experiment 2
Preincubation 37 °C, incubation without inhibitor	0.65 ± 0.065	0.48 ± 0.035
Preincubation 37 °C, incubation with inhibitor	0.45 ± 0.043	0.30 ± 0.015
Preincubation 0 °C, incubation without inhibitor	0.63 ± 0.057	0.51 ± 0.066
Preincubation 0 °C, incubation with inhibitor	0.54 ± 0.054	0.33 ± 0.031
$D_{0.05}$	0.067	0.084
$D_{0.001}$	0.121	0.152

Guinea-pig intestinal rings were preincubated in Krebs bicarbonate buffer for 30 min in melting ice, then rewarmed for 5 min at 37 °C in a potassium-free buffer before the determination of the influx of the radioactive substrate by means of a further incubation of 2 min, also in the absence of potassium ions. Control tissues were preincubated for 30 min at 37 °C and then subjected to the same incubations in the absence of potassium. In Experiment 1, the substrate was 1 mM ^{14}C-L-phenylalanine and the inhibitor 10 mM D-galactose, whereas in Experiment 2, the substrate was 1 mM ^{14}C-β-methyl-D-glucoside and the inhibitor 10 mM L-phenylalanine. The results are the means ± SEM of nine experiments in each case and are expressed together with the least significant difference at two different probability levels, calculated from a random-block analysis of variance (*Lison*, 1968).

the strongest inhibitors of amino acid influx. We have accordingly examined the inhibition of phenylalanine influx provoked by two monosaccharides, galactose and β-methyl-glucoside, whose own transport properties differ considerably. As seen in Table 2, the influx of β-methyl-glucoside is substantially greater than that of galactose. As a corollary, it stimulates to a significantly greater extent both the unidirectional influx of sodium into intestinal rings and the transmural potential difference across intestinal sheets. Thus it would be expected, in accordance with the "accelerated-efflux hypothesis", that β-methyl-glucoside would inhibit phenylalanine influx more powerfully than galactose. In fact, the reverse is true: Fig. 4 illustrates the inhibition of phenylalanine transport by these two sugars, and demonstrates that galactose is the more powerful inhibitor. This point has also been made incidentally by *Kimmich* and *Randles* (1973): These workers examined the interactions between valine and 3-O-methyl-glucose in the chicken intestine and found that the uptake of valine was inhibited to a much greater extent by 3-O-methyl-glucose than that of the sugar by the amino acid, although Na^+-dependent amino acid influx was substantially greater than that of the sugar. They also concluded that

Table 2. *Quantitative comparison of the co-transport of sodium ions with galactose or β-methyl-glucoside*

	Galactose	β-methyl-glucoside	P
Experiment 1:			
Influx of monosaccharide (μmol/g fresh tissue · 2 min)	0.25 ± 0.015	0.68 ± 0.021	< 0.001
Experiment 2:			
Stimulation of sodium influx (ΔJ_{Na}) in presence of sugar (μmol/g fresh tissue)	0.47 ± 0.166	0.92 ± 0.153	< 0.001
Experiment 3:			
Increase in transmural potential elicited by sugar (mV)	1.7 ± 0.17	2.9 ± 0.31	< 0.01

In Experiment 1, guinea-pig intestinal rings were incubated two min in a solution of 1 mM monosaccharide in Krebs bicarbonate buffer. The results are means ± SEM of five experiments. In Experiment 2, rings were incubated for 2 min in a ^{22}Na-labelled buffer containing 40 mM sodium ions. The sugars were added to appropriate flasks, and the increase in the influx of sodium ions in the presence of 5 mM sugar (ΔJ_{Na}) was assessed. The results are the means of 10 experiments. In Experiment 3, sheets of guinea-pig intestine were clamped between flux chambers and the stimulation of the transmural potential in the presence of 7 mM sugar was determined. The results are the means of 8 experiments. In each case, the statistical probability (P) was evaluated by paired-data analysis.

discharge of the sodium gradient, as envisaged by *Semenza* (1971), could not be responsible for heterologous interactions.

Conclusions

The present experiments confirm that, in the guinea-pig ileum, tryptophan is transported by the same site as phenylalanine, and most probably only one saturable mechanism is involved in the transfer of these two amino acids. This observation is important insofar as it validates the use of either of these amino acids as models for the exploration of kinetic relationships in the intestine.

The influx of both amino acids is inhibited by monosaccharides. In our studies of heterologous interactions, we have generally used phenylalanine as the model amino acid, and galactose and β-methyl-glucoside as the inhibitory sugars. This experimental design has permitted us to test the validity of two different hypotheses concerned with the interactions between sugars and amino acids for transfer across the brush-border membrane. Since the rates of transfer of the two sugars are very different, galactose being transported much more slowly than β-methyl-glucoside, the fact that galactose is a more powerful inhibitor of amino acid influx than is β-methyl-glucoside is incompatible with any hypothesis that requires inhibitory potency to

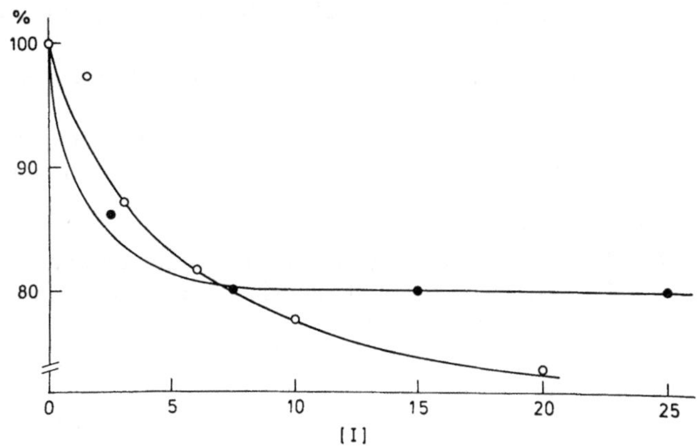

Fig. 4. Comparison of the inhibition of L-phenylalanine influx by different concentrations of D-galactose and β-methyl-D-glucoside. Intestinal rings were incubated for 2 min in a solution of 1 mM ^{14}C-L-phenylalanine in the presence of different concentrations of the sugars. The galactose series (open circles) represents the means of 7 experiments and the β-methyl-glucoside series (closed circles) the means of 13 experiments. For the purpose of the presentation, the values have been normalised and the uptake is expressed as the percentage of the respective controls

be correlated with rates of transport. The observation is, however, consistent with the "allosteric-interaction hypothesis" which assumes that the membrane matrix contains a series of distinct binding sites for different types of substrates. These binding sites, although different, are not entirely independent, but are functionally related and may interact in an allosteric manner. This model is indeed amenable to quantitative analysis, and the predictions and quantitative confirmation of its postulates, with the aid of experiments along the lines of that illustrated in Fig. 4, will be the subject of a future paper.

Acknowledgements

We are grateful to Francisco Sepúlveda (Cambridge) for access to data prior to publication. We are indebted to Pierrette Ganguillet, Sylvianne Koenig and Dominique Mettraux for technical assistance, and to Winefred Happee for secretarial aid. The programme received financial support from the Fonds National Suisse de la Recherche Scientifique (contract 3.014-0.76), the I.N.S.E.R.M., France (contract 78.4.372 AU) and the D.G.R.S.T., France (contract 76.7.1170).

References

Alvarado, F.: Transport of sugars and amino acids in the intestine: Evidence for a common carrier. Science *151*, 1010—1013 (1966).

Alvarado, F.: Hypothesis for the interaction of phlorizin and phloretin with membrane carriers for sugars. Biochim. biophys. Acta *135*, 483—495 (1967).

Alvarado, F.: Amino acid transport in hamster small intestine: Site of inhibition by D-galactose. Nature *219*, 276—277 (1968).

Alvarado, F.: Interrelation of transport systems for sugars and amino acids in small intestine. In: Intestinal Transport of Electrolytes, Amino Acids and Sugars (*Armstrong, W. McD., Nunn, A. S.*, eds.), pp. 281—315. Springfield: Ch. C Thomas. 1971.

Alvarado, F., Mahmood, A.: Cotransport of organic solutes and sodium ions in the small intestine: a general model. Amino acid transport. Biochemistry *13*, 2882—2890 (1974).

Chez, R. A., Schultz, S. G., Curran, P. F.: Effect of sugars on transport of alanine in intestine. Science *153*, 1012—1013 (1966).

Cleland, W. W.: The statistical analysis of enzyme kinetic data. Adv. Enzymol. *29*, 1—32 (1967).

Curran, P. F., Schultz, S. G., Chez, R. A., Fuisz, R. E.: Kinetic relations of the Na-amino acid interaction at the mucosal border of the intestine. J. gen. Physiol. *50*, 1261—1286 (1967).

Frizzell, R. A., Schultz, S. G.: Distinction between galactose and phenylalanine effects on alanine transport in rabbit ileum. Biochim. biophys. Acta *233*, 485—488 (1971).

Inui, Y., Christensen, H. N.: Discrimination of single transport systems. The Na^+-sensitive transport of neutral amino acids in the Ehrlich cell. J. gen. Physiol. *50,* 203—224 (1966).
Kimmich, G. A., Randles, J.: Interaction between Na^+-dependent transport systems for sugars and amino acids. Evidence against a role for the sodium gradient. J. memb. Biol. *12,* 47—68 (1973).
Lison, L.: Statistique appliquée à la biologie expérimentale, p. 169. Paris: Gauthier-Villars. 1968.
Munck, B. G., Rasmussen, S. N.: Characteristics of rat jejunal transport of tryptophan. Biochim. biophys. Acta *389,* 261—280 (1975).
Murer, H., Sigrist-Nelson, K., Hopfer, U.: On the mechanism of sugar and amino acid interaction in intestinal transport. J. biol. Chem. *250,* 7392—7396 (1975).
Neame, K. D., Richards, T. G.: Elementary kinetics of membrane carrier transport, p. 79. Oxford: Blackwell. 1972.
Newey, H., Smyth, D. H.: Effects of sugars on intestinal transfer of amino acids. Nature *202,* 400—401 (1964).
Robinson, J. W. L.: Role of intracellular ion concentrations on non-electrolyte influx. Gastroent. clin. biol. *2,* 337—338 (1978).
Robinson, J. W. L., Alvarado, F.: Interaction between the sugar and amino acid transport systems at the small intestinal brush border: A comparative study. Pflügers Arch.-Europ. J. Physiol. *326,* 48—75 (1971).
Robinson, J. W. L., Alvarado, F.: Comparative aspects of the interactions between sugar and amino acid transport systems. In: Intestinal Permeation (*Kramer, M., Lauterbach, F.,* eds.), pp. 145—162. Amsterdam: Excerpta Medica. 1977.
Robinson, J. W. L., Felber, J.-P.: Compartments of the uptake of amino acids by intestinal fragments during in vitro incubation. Gastroenterologia (Basel) *104,* 335—342 (1965).
Robinson, J. W. L., Sepúlveda, F. V.: Kinetics of sodium-activated phenylalanine influx in the guinea-pig intestine in vitro. J. Physiol. (London) *266,* 42P—43P (1977).
Semenza, G.: On the mechanism of mutual inhibition among sodium-dependent transport systems in the small intestine: A hypothesis. Biochim. biophys. Acta *241,* 637—649 (1971).
Sepúlveda, F. V., Smith, M. W.: Discrimination between different entry mechanisms for neutral amino acids in rabbit ileal mucosa. J. Physiol. (London) *282,* 73—90 (1978).

Authors' address: Dr. *J. W. L. Robinson,* Département de Chirurgie Expérimentale, Centre Hospitalier Universitaire Vaudois, CH-1011 Lausanne, Switzerland.

Journal of Neural Transmission, Suppl. 15, 139—151 (1979)

Transport of Tryptophan and Other Amino Acids by Mammalian Erythrocytes

J. D. Young and **J. C. Ellory**

ARC Institute of Animal Physiology, Cambridge, and Physiological Laboratory, Cambridge, U.K.

With 4 Figures

Summary

Mammalian erythrocytes are known to have 4 distinct amino acid transport systems. The C system is found in sheep erythrocytes and is specific for neutral amino acids of intermediate size. Human cells lack the C system and have two alternative neutral amino acid transport mechanisms: the L system which is selective for large hydrophobic amino acids and a Na-dependent alanine-cysteine system. The L system is absent from ruminant and cat erythrocytes. Human, rabbit and cat cells, but not sheep erythrocytes, have a specific dibasic amino acid transport mechanism (Ly system). L-Tryptophan uptake by human erythrocytes is mediated by two routes: a saturable high-affinity transport mechanism selective for L-tryptophan and a low affinity uptake mediated by the L system. The saturable component of L-tryptophan transport occurs by a previously unidentified uptake system, and is the major route for L-tryptophan uptake at physiological substrate concentrations. This system is absent from sheep and cat erythrocytes.

Introduction

The mammalian erythrocyte has received considerable attention in membrane transport studies. However, information on the permeability of erythrocytes to amino acids is surprisingly limited and it is only within the last few years that the subject has been investigated in any detail. It is now established that these cells possess a number of different amino acid transport systems which play an important role in normal erythrocyte function. In the present paper we summarize the properties of these transport systems and in addition present previously unpublished data on L-tryptophan transport into mammalian erythrocytes.

Erythrocyte Amino Acid Transport Systems

The C System

The first evidence for the existence of the C or cysteine transport system came from an investigation of the amino acid permeability characteristics of GSH-deficient sheep erythrocytes. *Tucker* and *Kilgour* (1970) described an inherited GSH-deficiency in certain animals of the Finnish Landrace breed. This deficiency, which is inherited in a simple autosomal recessive manner, results in a markedly diminished erythrocyte life-span and an increased susceptibility to oxidative stress (*Tucker*, 1974; *Tucker* and *Kilgour*, 1973). The lesion was subsequently shown to be associated with high intracellular concentrations of a number of amino acids notably ornithine and lysine (*Ellory et al.*, 1972) and amino acid uptake measurements revealed that the cells had a markedly diminished permeability to a number of amino acids particularly L-alanine and L-cysteine (*Young et al.*, 1975). A comparison of the amino acid transport characteristics of normal and GSH-deficient erythrocytes established that normal sheep erythrocytes possess a stereospecific transport system for neutral amino acids of intermediate size which is inoperative in GSH-deficient cells (*Young et al.*, 1976). The functional absence of this transport system (the C or cysteine system) from GSH-deficient erythrocytes was not a secondary consequence of low intracellular GSH or high ornithine and lysine concentrations but rather represents a membrane transport defect.

The structural specificity of the C system was characterized in detail by competition and accelerative exchange diffusion studies (*Young* and *Ellory*, 1977 a). These data together with the permeability differences between normal and GSH-deficient cells permitted the structural selectivity of the system to be established in detail. The optimal substrates are small neutral amino acids of intermediate size such as L-alanine, L-α-amino-n-butyrate, L-cysteine and L-norvaline. Decreasing or increasing the length of the amino acid side chain, branching of the side chain or the presence of an aromatic ring all result in a marked decrease in affinity. D-amino acids and α-amino-isobutyrate do not interact with the system indicating strict structural requirements at the α-carbon atom which include the presence of both the α-amino and carboxyl groups. In addition the transport system has the ability to distinguish between L-cysteine and L-serine on the one hand and L-methionine and L-norleucine on the other demonstrating a high degree of selectivity at two positions along the amino acid side chain. In contrast, the system accepts the presence of a terminal amino group on the side chain although the bulkier guanidinium group of L-arginine is excluded.

The specificity of the C system, which appears to be the major route for amino acid transport in sheep erythrocytes, resembles that of the ASC system which has been found in a number of cell types including rabbit reticulocytes. However, the ASC system unlike the C system is Na-dependent. The structural selectivity of the C system seems to be directed towards the transport of L-cysteine for intracellular GSH biosynthesis and its absence from GSH-deficient cells is presumably responsible for the low intracellular GSH concentration. The absence of the transport mechanism may also account for the high concentration of dibasic amino acids in these cells, since the C system is the major route for dibasic amino acid transport in normal sheep erythrocytes (*Young et al.*, 1976; *Young et al.*, 1978). The origin of the amino acids which accumulate in GSH-deficient cells remains to be elucidated. One possibility is that they arise from protein degradation during reticulocyte maturation (*Tucker et al.*, 1977; *Young* and *Ellory*, 1977 b).

The importance of the C system to the metabolism and viability of the sheep erythrocyte raises the question of whether the same system is present in other mammalian species. Although there are large differences between species with regard to the uptake of certain amino acids (*Young* and *Ellory*, 1977 b), it is striking that the uptake of L-alanine is relatively uniform and if the stereospecific alanine flux (*i.e.* the difference between L- and D-alanine uptake rates) is considered, the uptake rate becomes the same for all species studied. These data are consistent with the C system being present in all erythrocytes, but with additional components present in some species. However, kinetic studies have demonstrated that the C system is not present in human erythrocytes. L-alanine and L-cysteine uptake by human cells are mediated by a Na-dependent alanine-cysteine system and by the L system (*Winter* and *Christensen*, 1964; *Ellory* and *Young*, 1977; *Ellory* and *Young*, 1978; *Young et al.*, 1978), neither of which occur in sheep erythrocytes. Dibasic amino acid transport in human erythrocytes is mediated by an uptake mechanism (Ly system) selective for this class of amino acids (*Gardner* and *Levy*, 1972; *Young et al.*, 1978). The Ly system is not present in sheep cells.

The L System

Winter and *Christensen* (1964) proposed that human erythrocytes possess a transport system (the L system) selective for amino acids with large hydrophobic side chains. Subsequent studies have verified this conclusion although the later investigations have found higher values for the kinetic constants (*i.e.* K_m and V_{max}) (*Hoare*, 1972 a, b; *Ellory* and *Young*, 1977; *Young* and *Ellory*, 1977 b; *Young et al.*, 1978). The L system is not Na-dependent and is only partially stereospecific with the apparent K_m value for L-leucine 4-fold lower than for the D-isomer. Although the optimal substrates are large neutral amino acids such as L-leucine and L-phenylalanine, the system has a significant affinity for smaller amino acids such as L-alanine and L-cysteine and approximately 50 % of the uptake of these amino acids at physiological substrate concentrations occurs by this route.

Hoare (1972 a, b) investigated the transport kinetics of this system in detail. The concentration dependence of influx and efflux were consistent with Michaelis-Menten kinetics and a simple model involving translocation of a membrane carrier, either alone or complexed with amino acid was proposed. *Hoare* found that carrier re-orientation was the rate-limiting step of L-leucine translocation, and binding of amino acid to the carrier markedly increased the re-orientation rate. The apparent K_m and V_{max} values of L-leucine influx and efflux were markedly temperature dependent and although the kinetic parameters of influx and efflux were similar at 37 °C, a marked kinetic asymmetry developed at low temperatures. Although not investigated in such detail, the kinetic characteristics of the C system are superficially similar.

The L system is also present in rat and rabbit erythrocytes, but with higher activity. In marked contrast, the system is absent from both ruminant and cat erythrocytes (*Young* and *Ellory*, 1977 b; *Young et al.*, 1978).

The Na-Dependent Alanine-Cysteine System

The early work of *Yunis* and *Arimura* (1965) showed a Na-dependent transport component for glycine and L-alanine in human erythrocytes. A detailed investigation of the concentration dependence of alanine uptake in either Na or choline medium revealed the presence of a high affinity, low capacity, Na-dependent transport system which was highly stereospecific (*Ellory* and *Young*, 1978; *Young et al.*, 1978). L-cysteine also showed a high affinity Na-dependent uptake component. Transport of L-cysteine by this route was readily inhibited by L-alanine suggesting that the two amino acids share a common entry mechanism.

This Na-dependent uptake route is not the result of white cell, platelet or reticulocyte contamination of the cell preparation and replacement of Na with cations other than choline demonstrated that the diminished uptake in the presence of choline was a true Na-replacement response. The substrate specificity of this uptake route has not as yet been investigated in detail. However, it may be analogous to the Na-dependent ASC system which is present in high activity in reticulocytes (*Thomas* and *Christensen,* 1970). At physiological substrate concentrations this transport system accounts for approximately half of the L-cysteine uptake into human cells. This uptake mechanism is not present in sheep erythrocytes.

The Ly System

Gardner and *Levy* (1972) suggested that human erythrocytes have a high affinity transport system for dibasic amino acids. The usefulness of this study was limited because no attempt was made to relate dibasic amino acid transport to the transport of other amino acids and it was therefore not clear whether, as in normal sheep cells, these amino acids share one of the neutral amino acid systems. The transport of dibasic amino acids by rabbit erythrocytes was investigated by *Christensen* and *Antonioli* (1969) and *Antonioli* and *Christensen* (1969). On the basis of competition and reticulocyte maturation studies it was proposed that these cells possess a transport system selective for dibasic amino acids (the Ly system).

Investigations in this laboratory have confirmed the existence of the Ly system in human erythrocytes and have also shown it to be present in cat erythrocytes but absent from sheep cells where the major route for dibasic amino acid transport is by the C system. An important difference between the dibasic amino acid specificities of the Ly and C systems is that the Ly system transports L-arginine. The Ly system is stereospecific and Na-independent (*Young et al.,* 1976; *Young* and *Ellory,* 1977 a; *Young et al.,* 1978). Although the C system has a much lower affinity for dibasic amino acids than the Ly system, it has a considerably higher V_{max}. The properties of the amino acid transport systems discussed in this section are summarized in Table 1.

Tryptophan Transport

Fig. 1 shows the concentration dependence of L-tryptophan uptake by human erythrocytes at 37 °C over the concentration range 1 μM to

Table 1. *Erythrocyte amino acid transport systems*

System	Specificity	Kinetic parameters: Amino acid	Apparent K_m (mM)	V_{max} (mmoles/l cells per hour)	Species distribution	Na-Dependence
C	Small neutral amino acids of intermediate size. Significant affinity for dibasic amino acids. Stereospecific	L-alanine L-cysteine L-lysine	19 14 $>$ 50	10 9 —	Positively identified only in sheep. May be present in all ruminants and some other species. Not present in human cells	No
L	Large neutral amino acids but with low affinity for small neutral amino acids. Partially stereospecific	L-leucine L-alanine L-cysteine	4.8 $>$ 50 $>$ 50	68 — —	Present in human, rabbit and rat cells. Not found in ruminant or cat erythrocytes	No
Na-dependent alanine-cysteine	L-alanine, L-cysteine, possibly glycine, stereospecific	L-cysteine	0.02	0.03	Present in human but not sheep erythrocytes	Yes
Ly	Dibasic amino acids, stereospecific	L-lysine L-arginine	0.03 0.09	0.44 0.40	Found in human, rabbit and cat erythrocytes. Not present in sheep cells	No

References to particular transport systems are given in the text. The kinetic constants refer to human and sheep erythrocytes.

50 mM. The initial rate of uptake, which was not Na-dependent, was linear over the low concentration range (1—22 μM). The concentration dependence at higher concentrations (0.1—5 mM and 1—50 mM) suggests the presence of two components of uptake: a saturable component with apparent K_m and V_{max} values of approximately 1.3 mM and 7.2 mmoles/l · cells per hour respectively and a high capacity nonsaturable component (10 mmoles/l · cells per hour at 25 mM). Fig. 2 shows the effect of 5—50 mM L-leucine on the initial rate of L-tryptophan uptake at 3.3 mM. Although L-leucine was an effective inhibitor of L-tryptophan influx, the inhibition was only partial ($\simeq$50 %) even at high inhibitor concentrations. Fig. 3 compares the concentration dependence of L-tryptophan uptake in the presence and absence of 50 mM L-leucine. The L-tryptophan curve in the absence of L-leucine is resolved into saturable and nonsaturable components. The results clearly demonstrate that L-leucine inhibition of L-tryptophan influx is associated with an almost complete loss of the nonsaturable component of L-tryptophan transport. In contrast L-leucine had little effect on the saturable uptake route. Table 2 summarizes the ability of a series of amino acids (10 mM) to inhibit L-tryptophan transport at a substrate concentration (0.2 mM) where most of the influx (approximately 85 %) occurs by the saturable

Table 2. *Effects of amino acids on L-tryptophan uptake by human erythrocytes*

Amino acid (10 mM)	L-tryptophan uptake (0.2 mM) % Control value
Glycine	100
Proline	96
L-alanine	102
L-α-amino-*n*-butyrate	96
L-valine	88
L-norleucine	82
L-leucine	82
L-phenylalanine	38
L-histidine	96
L-lysine	96
5-hydroxytryptamine	79

Values are the means of duplicate estimates. At this L-tryptophan concentration approximately 85 % of the uptake occurs by the saturable route. See legend to Fig. 1 for experimental details.

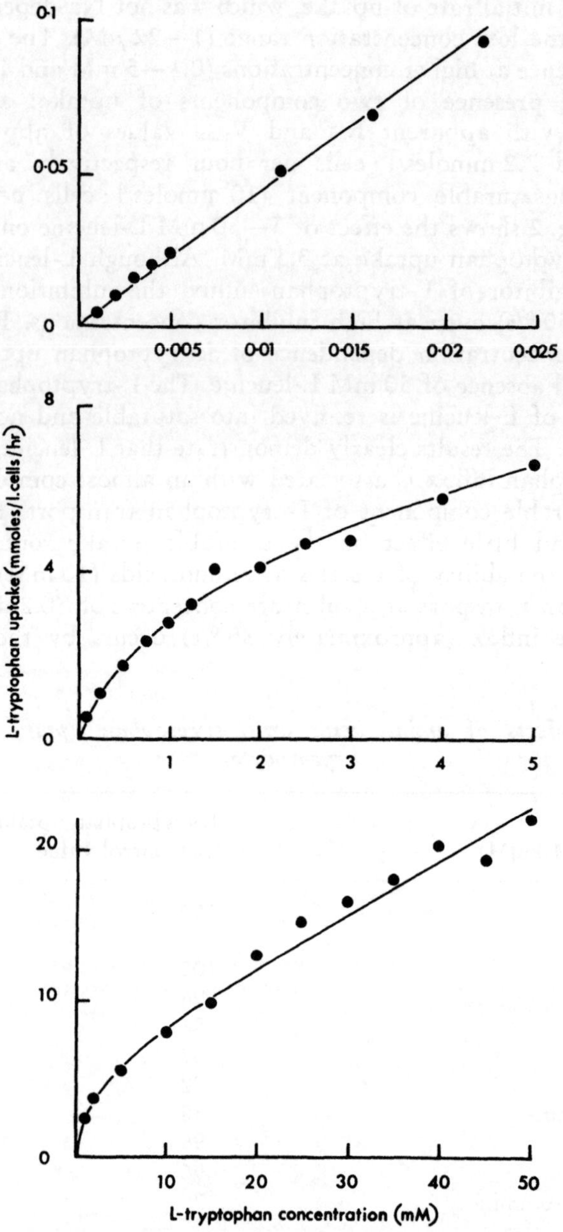

Fig. 1. Concentration dependence of L-tryptophan uptake by human erythrocytes. The initial rate of L-[methylene-^{14}C]-tryptophan uptake was measured at 37 °C as previously described (*Young et al.*, 1976). Iso-osmolality was maintained by adjusting the concentration of NaCl in the medium

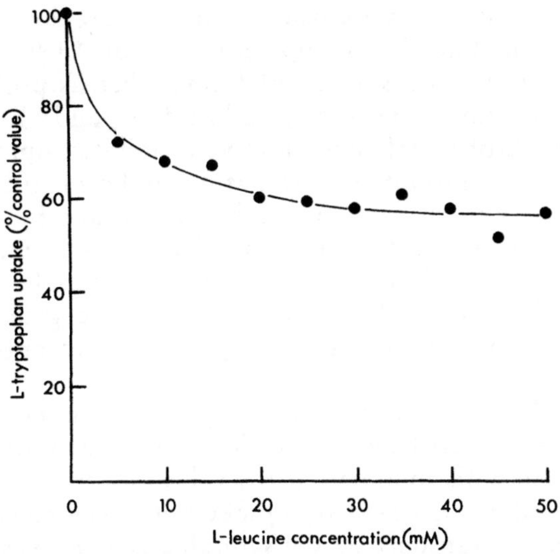

Fig. 2. Effect of L-leucine on L-tryptophan uptake by human erythrocytes. See legend to Fig. 1 for experimental details. The L-tryptophan concentration was 3.3 mM

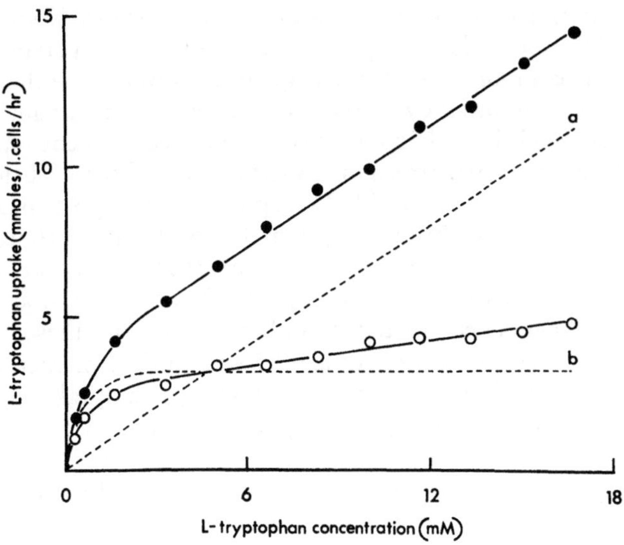

Fig. 3. Concentration dependence of L-tryptophan uptake by human erythrocytes in the presence and absence of L-leucine.

The L-leucine concentration was 50 mM (○). L-Tryptophan transport in the absence of added L-leucine (●) is resolved into linear (a) and saturable (b) components. See legend to Fig. 1 for other experimental details

uptake component. In agreement with the previous experiment, L-leucine only inhibited L-tryptophan uptake by 18 %. L-norleucine and L-valine produced a similar inhibition whereas proline, glycine, L-alanine, L-α-amino-n-butyrate, L-histidine and L-lysine were without effect. L-phenylalanine, however, inhibited uptake by 62 % suggesting that the saturable route has a significant affinity for this amino acid. 5-Hydroxytryptamine inhibited uptake by 21 %. In an other experiment the ability of L-tryptophan (2.5—20 mM) to inhibit L-leucine influx (2 mM) was investigated. L-tryptophan was found to be a poor inhibitor of L-leucine uptake with the degree of inhibition directly proportional to the L-tryptophan concentration (21 % at 20 mM).

These results demonstrate that L-tryptophan uptake by human erythrocytes is mediated by two distinct routes: a medium capacity saturable system with an apparent K_m value of approximately 1 mM and a nonsaturable, high capacity uptake route. The kinetic constants for L-tryptophan transport by the saturable system, particularly the V_{max}, are considerably lower than the corresponding constants for L-leucine transport in human cells (Table 1), and the L-leucine and L-tryptophan inhibition data demonstrate that L-tryptophan uptake by this route is *not* mediated by the L system. Instead the results suggest that transport occurs by another route which is selective for L-tryptophan, but which may also transport L-phenylalanine. Since the saturable component of L-tryptophan transport is neither sodium-dependent nor inhibited by L-alanine or L-lysine (the alanine-cysteine and Ly systems) it is probable that this uptake component represents a previously unidentified transport system. The ability of L-leucine to inhibit most of the nonsaturable L-tryptophan transport suggests that this uptake is mediated by the L system, but with a low affinity.

Unlike human erythrocytes, normal sheep and cat cells are relatively impermeable to L-tryptophan. Fig. 4 shows that the concentration dependence of L-tryptophan uptake in these species is linear over the concentration range 0.5—25 mM, demonstrating that sheep and cat cells lack the saturable L-tryptophan uptake mechanism. Furthermore the linear L-tryptophan uptake in these species (2.7 mmoles/l · cells per hour at 25 mM for both cell types) is 4-fold lower than the nonsaturable uptake component found in human erythrocytes. This correlates with the absence of the L system from sheep and cat cells (*Young* and *Ellory*, 1977 a; *Young et al.*, 1978).

Fig. 4 shows a further species variation in L-tryptophan transport. Uptake by rat erythrocytes is considerably faster than in human cells, and transport saturates at L-tryptophan concentrations above 20 mM. The mechanism(s) responsible for this uptake remain to be resolved.

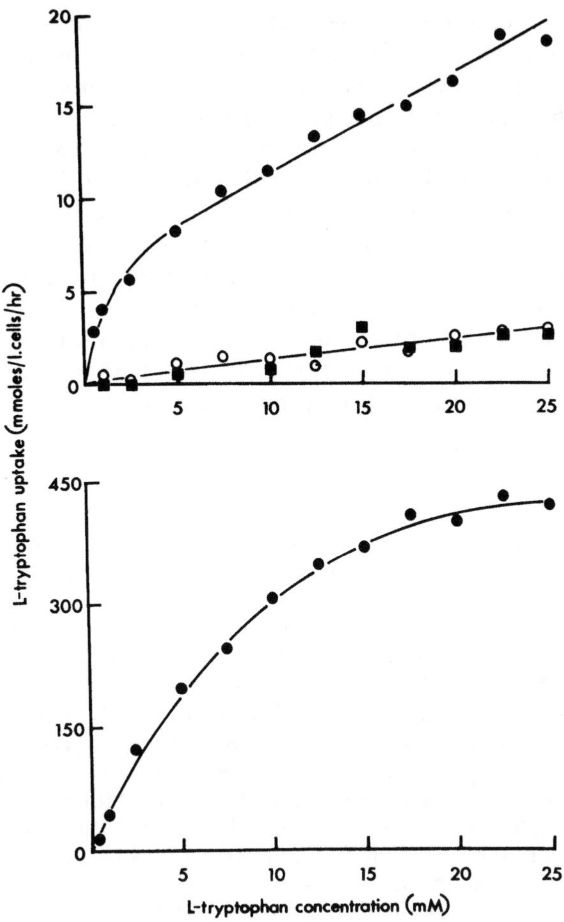

Fig. 4. Concentration dependence of L-tryptophan uptake by erythrocytes from different mammalian species.
See legend to Fig. 1 for experimental details. Human ●, sheep ○ and cat ■ erythrocytes, upper figure. Rat ● erythrocytes, lower figure

Conclusions

It is now established that mammalian erythrocytes possess a number of distinct amino acid transport systems with differing specificities, translocation capacities and species distributions. These systems are important for normal erythrocyte function and viability, but specific roles are not restricted to particular transport systems. Thus L-cysteine transport for intracellular GSH biosynthesis is mediated by the C system in sheep erythrocytes whereas the L and

Na-dependent alanine-cysteine systems combine to fulfil the same function in human cells. Similarly, dibasic amino acid transport in sheep erythrocytes occurs by the C system whereas other species have a specific dibasic amino acid transport system.

The data presented in this paper demonstrate that L-tryptophan uptake by human erythrocytes is mediated by two distinct routes: a high affinity saturable, sodium-independent uptake mechanism selective for L-tryptophan, and a low affinity uptake mediated by the L system. The saturable component of L-tryptophan transport occurs by a previously unidentified uptake system (see Table 1) and is the major route for L-tryptophan uptake at physiological substrate concentrations. Sheep and cat erythrocytes lack both transport systems and consequently transport L-tryptophan slowly.

References

Antonioli, J. A., Christensen, H. N.: Differences in the schedules of regression of transport systems during reticulocyte maturation. J. Biol. Chem. *244*, 1505—1509 (1969).

Christensen, H. N., Antonioli, J. A.: Cationic amino acid transport in the rabbit reticulocyte. J. Biol. Chem. *244*, 1497—1504 (1969).

Ellory, J. C., Young, J. D.: Neutral amino acid transport in erythrocytes from different mammalian species. J. Physiol. *272*, 43—44P (1977).

Ellory, J. C., Young, J. D.: Sodium-dependent amino acid transport in human erythrocytes. J. Physiol. *285*, 51—52P (1978).

Ellory, J. C., Tucker, E. M., Deverson, E. V.: Identification of ornithine and lysine at high concentrations in the red cells of sheep with an inherited deficiency of GSH. Biochim. Biophys. Acta *279*, 481—483 (1972).

Gardner, J. D., Levy, A. G.: Transport of dibasic amino acids by human erythrocytes. Metab. Clin. Exp. *21*, 413—431 (1972).

Hoare, D. G.: The transport of L-leucine in human erythrocytes: a new kinetic analysis. J. Physiol. *221*, 311—329 (1972 a).

Hoare, D. G.: The temperature dependence of the transport of L-leucine in human erythrocytes. J. Physiol. *221*, 331—348 (1972 b).

Thomas, E. L., Christensen, H. N.: Indications of spatial relations among structures recognizing amino acids at a transport receptor site. Biochem. Biophys. Res. Commun. *40*, 277—283 (1970).

Tucker, E. M.: A shortened lifespan of sheep red cells with a glutathione deficiency. Res. Vet. Sci. *16*, 19—22 (1974).

Tucker, E. M., Kilgour, L.: An inherited glutathione deficiency and a concomitant reduction in potassium concentration in sheep red cells. Experientia *26*, 203—204 (1970).

Tucker, E. M., Kilgour, L.: The effect of anaemia on sheep with inherited differences in red cell reduced glutathione (GSH) concentrations. Res. Vet. Sci. *14*, 306—311 (1973).

Tucker, E. M., Wright, P. C., Young, J. D.: Influence of arginase deficiency on amino acid concentrations in sheep erythrocytes with a normal and with a defective transport system for amino acids. J. Physiol. *271,* 47—48P (1977).

Winter, C. G., Christensen, H. N.: Migration of amino acids across the membrane of the human erythrocyte. J. Biol. Chem. *239,* 872—878 (1964).

Young, J. D., Ellory, J. C.: Substrate specificity of amino acid transport in sheep erythrocytes. Biochem. J. *162,* 33—38 (1977 a).

Young, J. D., Ellory, J. C.: Red cell amino acid transport. In: Membrane Transport in Red Cells (*Ellory, J. C., Lew, V. L.,* eds.), pp. 301—325. New York-London: Academic Press. 1977 b.

Young, J. D., Ellory, J. C., Tucker, E. M.: Amino acid transport defect in glutathione-deficient sheep erythrocytes. Nature *254,* 156—157 (1975).

Young, J. D., Ellory, J. C., Tucker, E. M.: Amino acid transport in normal and GSH-deficient sheep erythrocytes. Biochem. J. *154,* 43—48 (1976).

Young, J. D., Jones, S. E. M., Ellory, J. C.: Amino acid transport in mammalian erythrocytes (in preparation, 1978).

Yunis, A. A., Arimura, G. K.: Amino acid transport in blood cells. II. Patterns of transport of some amino acids in mammalian reticulocytes and mature red blood cells. J. Lab. Clin. Med. *66,* 177—186 (1965).

Authors' address: Dr. *J. D. Young,* ARC Institute of Animal Physiology, Babraham, Cambridge, U.K.

Journal of Neural Transmission, Suppl. 15, 153—160 (1979)

Zero-Trans Uptake of L-Tryptophan in the Human Erythrocyte

R. Rosenberg

Psychochemistry Institute, University of Copenhagen, Rigshospitalet, Copenhagen, Denmark

With 1 Figure

Summary

Some results from ongoing studies of the transport of some neutral amino acids across the human erythrocyte membrane (L-tryptophan and L-leucine) at 25 °C are presented.

The unidirectional flux into the cell at initial rate as a function of the substrate concentration can be described by a 3-parameter model (*i.e.* a carrier in parallel with an apparent leak component). From the estimated kinetic parameters an apparent membrane permeability of the carrier (substrate concentrations $S_1 \ll K^{zt}_{12}$) has been calculated. The permeability of L-tryptophan ($1.03 \cdot 10^{-8}$ cm/sec) was significantly lower than the permeability of L-leucine ($2.36 \cdot 10^{-7}$ cm/sec). The contribution of the leak permeability to the total permeability at low concentrations was about 10 % for L-tryptophan, but only 1 % for L-leucine.

From inhibition studies it is indicated that L-tryptophan passing the membrane via the apparent leak component is transported by the L-leucine carrier, the affinity being low.

The saturable component of L-tryptophan transport does not correspond to any known amino acid transport system, and it is thus strongly suggested that a new L-tryptophan specific transport system operating at physiological concentrations has been described.

Introduction

During the last few years an increasing number of studies have been published within biological psychiatry on various aspects of the functions of peripheral blood cells.

The human platelet has been suggested to serve as a model of synaptosomes (*Stahl*, 1977), the lymphocyte has been studied to demonstrate linkage of affective disorders to autosomal markers (*Shapiro et al.*, 1977), and the erythrocyte has been used in search of etiological genetic markers (*Gershon*, 1978; *Rafaelsen*, to be published).

Within the field of biophysics the erythrocyte is considered to be a very important cell, which can be used in studies of many basic features essential to biological transport functions.

From a clinical point of view the blood cells offer great advantages as being easily available.

This paper will present some results from ongoing studies of characterization of the transport of some neutral amino acids (L-tryptophan, L-leucine and L-phenylalanine) across the membrane of the human erythrocyte, focusing on the zero-trans uptake of L-tryptophan at 25 °C, pH = 7.4.

Experimental Procedure

Freshly drawn, heparinized human blood was washed three times with isotonic saline (pH = 7.4), and the buffy coat was sucked off.

For the study of efflux and the cellular concentration at equilibrium, cells were loaded by incubation at 25 °C with appropriate medium (^{14}C-labelled L-tryptophan and L-leucine [New England Nuclear Corp.]) at a length of time sufficient to reach equilibrium.

In efflux experiments packed cells were injected into the medium during vigorous stirring. The cells were separated from the medium at fixed time intervals by centrifugation, a part of the supernatant was removed and transferred to counting in a scintillation spectrometer.

For the study of influx packed cells were transferred to buffered saline containing the labelled amino acid during vigorous stirring in a test tube placed in a waterbath (hematocrit was about 5 %). The cells were separated from the medium by centrifugation through di-n-butylphthalate, and washed with a stopper solution. The cells were then lysed, the hemoglobin absorbance of the hemolysate was determined, trichloroacetic acid was added, and after centrifugation the supernatant was transferred for counting.

The common approach was to follow isotope concentration towards equilibrium as a function of time. The apparent membrane permeability P (cm/sec) is defined as

$$k = (PA)\,(1/V_1 + 1/V_2)$$

where k is the rate coefficient of tracer flux, A the volume independent surface area of the membrane, V_1 and V_2 the volume of the extracellular and intracellular compartment.

The unidirectional flux at initial rate U_{12} (mmol/cm^2 · sec) is given by

$$U_{12} = P\,S$$

where S is the extracellular concentration of L-tryptophan (zero-trans uptake experiments : mol/l).

When the volume of the cellular compartment and the membrane area are known, the flux can be expressed as mmol/l cell water · min.

The nomenclature follows *Lieb* and *Stein* (1974).

More details especially of the various experimental techniques will be presented in a forthcoming paper (*Rosenberg*, to be published) including the results from the L-leucine transport study mentioned in the present paper.

Results

When erythrocytes are incubated in media containing different concentrations of L-tryptophan, an apparent concentration gradient across the membrane was found at equilibrium all over the concentration range studied (0.025—25 mM). The cellular concentration (mmol/l cell water) was 1.75 (mean) times higher than the concentration in the medium. No gradient for L-leucine was found.

The magnitude of the gradient was independent of incubation for 48 hours without glucose in the medium, and no significant amount of metabolism was found.

The time course for tracer efflux (equilibrium exchange) at 0.8 mM L-tryptophan was well described by a two compartment model with constant volumes and permeability suggesting that the apparent higher cellular concentration is caused by binding of L-tryptophan to cellular components.

The initial rate of the flux of L-tryptophan (U_{12}) into the cell was measured as a function of the extracellular concentration of L-tryptophan (S_1) in the range 0.3—40 mM ("zero-trans influx").

The results are presented in Fig. 1 (upper curve). A simple carrier model without the presence of unstirred layer predicts a "Michaelis-Menten" relationship between flux and substrate concentration (*i.e.* a 2-parameter model), but a more complicated model is suggested from the curve.

In Table 1 is shown the results from a non-linear least square regression analysis of initial uptake rate vs. substrate concentration using a 3-parameter model (*i.e.* a carrier in parallel with an apparent leak):

$$U_{12} = \left(V^{zt}_{12}\,S_1\right) / \left(K^{zt}_{12} + S_1\right) + K_D\,S_1$$

where U_{12} is the unidirectional flux (mmol/l · min), V^{zt}_{12} the maximum velocity (mmol/l · min), K^{zt}_{12} the concentration of half-maximum

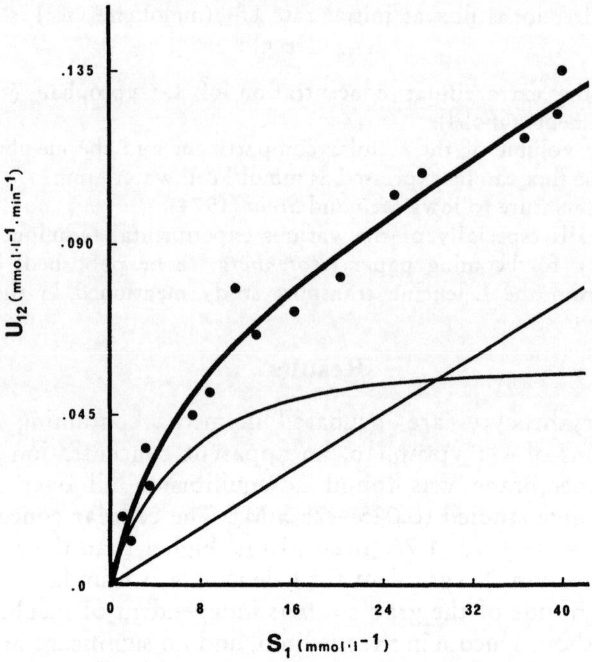

Fig. 1. Zero-trans influx of L-tryptophan (25 °C, pH = 7.4) as described under experimental procedure (upper curve). The ordinate gives the initial rate of the unidirectional flux (mmol/l cell water · min). The abscissa gives the concentration of L-tryptophan (range: 0.3—40 mM). The two lower curves are predicted from a non-linear least square regression analysis assuming a 3-parameter model (see text)

Table 1. *Estimation of the kinetic parameters from zero-trans influx experiments of L-tryptophan and L-leucine by non-linear least square regression of a 3-parameter mode (model I) and a 2-parameter model (model II) (see text)*

Models

$$\text{I: } \frac{VS}{K+S} + K_D S \qquad \text{II: } \frac{VS}{K+S}$$

	Model	V_{12}^{zt} (mM · min⁻¹)	K_{12}^{zt} (mM)	K_D (min⁻¹)
L-Trp:	I	0.059 ± 0.011	3.53 ± 1.31	0.0019 ± 0.0003
L-Leu:	I	0.604 ± 0.072	1.60 ± 0.50	0.0046 ± 0.0020
	II	0.771 ± 0.041	2.78 ± 0.58	

velocity (mmol/l), and K_D a constant proportional to the permeability coefficient of the leak component (min^{-1}).

The two components (*i.e.* the carrier and the leak) are shown separately in Fig. 1 (lower curves), the leak being the linear one.

In Table 1 the estimated kinetic parameters of the L-tryptophan transport is compared to the parameters of L-leucine transport, which have been estimated assuming a 2- and a 3-parameter model (*Rosenberg*, in preparation). Both models fit the data equally well.

Comparing the 3-parameter solutions the K_{12}^{zt} and K_D are found to be of the same magnitude, whereas the V_{12}^{zt} of L-leucine transport is about 10 times greater than the corresponding value of L-tryptophan. This means that the apparent membrane permeability at low substrate concentration ($S_1 << K_{12}^{zt}$) of L-tryptophan ($1.03 \cdot 10^{-8}$ cm/sec) is about 20 times smaller than the permeability of L-leucine ($2.36 \cdot 10^{-7}$ cm/sec) at 25 °C, pH 7.4, considering the carrier component.

Furthermore the contribution of the leak permeability to the total permeability at low concentration of the substrate ($S_1 << K_{12}^{zt}$) is about 10 % for L-tryptophan, but only 1 % for L-leucine.

In order to study the substrate specificity of the saturable component some competition and trans-acceleration experiments were performed. The uptake of L-leucine (0.6 mM) and the efflux of L-leucine (intracellular concentration: 12 mM) were measured in the presence of a competing amino acid. The results (percentage of controls) are the means of duplicate estimates and are given in Table 2.

The presence of a "L-system" (transport system for some large neutral amino acids) is suggested as several other amino acids tested had no effect on the transport of L-leucine. L-tryptophan has however a low affinity to the system.

Table 2. *Inhibition of L-leucine (0.6 mM) uptake by competing amino acids (10 mM) and trans-acceleration of L-leucine (12 mM) efflux by the same amino acids (50 mM)*

Competing amino acid	% Inhibition	% Trans-acceleration
L-leucine	78	102
L-phenylalanine	69	98
L-methionine	46	86
L-valine	45	56
L-tryptophan	9	10

The inhibition effect of different amino acids at a high concentration (30 mM) on the uptake of L-tryptophan (0.12 mM) is shown in Table 3. At this concentration of L-tryptophan the contribution of the apparent leak component to the total membrane permeability is about 10 %. Of the amino acids tested only L-phenylalanine, L- and D-tryptophan produced a marked inhibition.

Furthermore the concentration dependency of the L-leucine inhibition of the L-tryptophan uptake was unexpected. The results from Table 4 strongly suggest that the apparent leak component is inhibited by L-leucine. The leak component thus represents low-affinity transport of L-tryptophan via the L-system.

Table 3. *Inhibition of L-tryptophan (0.12 mM) uptake by competing amino acids (30 mM)*

Competing amino acid	% Inhibition
L-leucine	35
L-phenylalanine	90
L-methionine	40
L-valine	23
L-tryptophan	92
D-tryptophan	85
L-alanine	0
L-glycine	8
L-proline	26
L-aspartic acid	15
L-lysine	7

Table 4. *Inhibition (percentage of controls) of L-tryptophan uptake (0.3, 10, 25 mM) by 15 mM L-leucine. The contribution of the non-saturable component to the total membrane permeability is given for comparison*

L-tryptophan (mM)	Non-saturable component (%)	% Inhibition
0.3	11	35
10	30	35
25	48	57

Discussion

The analysis of the zero-trans influx experiments of L-tryptophan and L-leucine at 25 °C, pH = 7.4 suggests that the kinetics can be described by a 3-parameter model, *i.e.* a saturable component in parallel with an unsaturable. This model has frequently been reported for the influx of amino acids into solid tissues in vitro and into suspensions of isolated cells (*Winter* and *Christensen*, 1964; *Cohen*, 1975).

A detailed kinetic analysis of the saturable component of L-leucine transport has been performed (*Rosenberg* and *Clemmensen*, 1977). The results were consistent with the simple carrier model of *Lieb* and *Stein* (1974). From the experiments of the substrate specificity and studies on pH-dependency, Na^+-dependency and exchange-property of the carrier mediated transport (*Rosenberg*, in preparation), it is indicated that a common transport system of the large neutral amino acids (L-leucine, L-methionine, L-phenylalanine and L-valine) exists in the membrane of the human erythrocyte. This system corresponds to some extent to the "L-system" of *Christensen* (1965), which has been described in several tissues and cells (*e.g.* the blood-brain barrier [*Yudilevich et al.*, 1972]). The low membrane permeability for L-tryptophan mediated through the L-system carrier is however a striking finding.

At physiological concentrations the main route of transport for L-tryptophan is via the saturable component. Although L-phenylalanine produced a pronounced inhibition of L-tryptophan uptake, L-phenylalanine is not necessarily also transported by this system. Results from a kinetic analysis of L-phenylalanine (*Rosenberg*, in preparation) indicate that the transport is mainly mediated by the L-system.

The small inhibition effect on the L-tryptophan uptake by different amino acids tested (*cf.* Table 3) thus strongly suggests that a new L-tryptophan specific transport system in the membrane of the human erythrocyte has been described. The results from the present study correspond to a large extent to the results of the study of *Young* and *Ellory* (this volume) presented at this symposium.

Finally, it should be mentioned that the concept of a "carrier" refers to a specific kinetic model (*Lieb* and *Stein*, 1974), and that the phenomenological rate equations derived from the model are but useful guides to which any postulated molecular mechanism of transport must conform.

Acknowledgements

This study was supported by grants from the Danish Medical Research Council, the NOVO fond, the Nordisk Insulinfond, and Dr. med. vet. Axel Thomsen and Hustru Martha Thomsen f. Haugen-Johansens legat.

References

Christensen, H. N.: Some special kinetic problems of transport. Adv. Enzym. *32,* 1—20 (1969).

Cohen, S. R.: A comparison of the rate equations, kinetic parameters, and activation energies for the initial uptake of L-lysine, L-valine, γ-aminobutyric acid, and α-aminoisobutyric acid by mouse brain slices. J. Membrane Biol. *22,* 53—72 (1975).

Gershon, E. S.: The search for genetic markers in affective disorders. In: Psychopharmacology, a Generation of Progress (*Lipton, M. A., DiMascio, A., Killam, K. F.,* eds.), 1st ed., pp. 1197—1213. New York: Raven Press. 1978.

Lieb, W. R., Stein, W. D.: Testing and characterizing the simple carrier. Biochim. Biophys. Acta *373,* 178—196 (1974).

Rafaelsen, O. J.: ABO and HLA antigens in affective disorders (in press).

Rosenberg, R., Clemmensen, S.: Leucine transport in the human erythrocyte. Abstract B 1—3, 301, 5/6/7. FEBS 11th meeting, Copenhagen, Denmark, 1977.

Shapiro, R. W., Ryder, L. P., Svejgaard, A., Rafaelsen, O. J.: HLA antigens and manic-depressive disorders: further evidence of an association. Psychol. Med. *7,* 387—396 (1977).

Stahl, S. S.: The human platelet. Arch. gen. Psychiat. *34,* 509—516 (1977).

Winter, C. G., Christensen, H. N.: Migration of amino acids across the membrane of the human erythrocyte. J. Biol. Chem. *239,* 872—878 (1964).

Yudilevich, D. L., De Rose, N., Sepúlveda, F. V.: Facilitated transport of amino acids through the blood-brain barrier of the dog studied in a single capillary circulation. Brain Res. *44,* 569—578 (1972).

Author's address: Dr. *R. Rosenberg,* Psychochemistry Institute, University of Copenhagen, Rigshospitalet, 9, Blegdamsvej, DK-2100 Copenhagen Ø, Denmark.

Journal of Neural Transmission, Suppl. 15, 161—163 (1979)

Tryptophan Accumulation by Blood Platelets of Depressed Patients

K. Wood, C. Swade, and **A. Coppen**

MRC Neuropsychiatry Laboratory, Epsom, Surrey, U.K.

Summary

The accumulation of tryptophan by blood platelets has been investigated in depressive patients and controls using short incubation times and low substrate concentrations. The accumulation of tryptophan by the platelet is significantly greater in the acutely depressed patients than the control group. The results are discussed with reference to tryptophan transport in depression and also to the regulation of plasma levels of tryptophan.

The blood platelet and central nervous system serotonergic synaptosomes share many morphological and biochemical characteristics (*Sneddon,* 1973; *Stahl,* 1977). The blood platelet has, therefore, been used as an easily obtainable model of 5-hydroxytryptamine (5-HT) neurones. Kinetic analysis using low substrate concentrations and measuring the initial uptake rate have shown that the uptake of 5-HT into platelets of depressed patients is decreased (*Coppen et al.*, 1978). In the present investigation we have studied the uptake of tryptophan, the precursor of 5-HT, into blood platelets of depressed patients and controls.

Male and female patients who were diagnosed to be suffering from primary depressive illness were studied; none had a history of mania. These patients were assessed for severity of depression by the first sixteen items of the Hamilton Rating Scale (HRS) (*Hamilton,* 1960). Only patients who scored sixteen or more on this scale after a 7—10 day drug-free assessment phase were investigated. Men and women who volunteered to act as control subjects who had no known history of psychiatric disorder nor had taken any medication 7—10 days prior to biochemical testing were also investigated. A blood sample was obtained by venepuncture with the minimum of stasis at approximately 9 a.m. (after overnight fasting) from control subjects and patients. The uptake of ^{14}C-labelled L-tryptophan into the platelets was determined using a method that will be described elsewhere (*Wood et al.*, unpublished results).

Patients who subsequently recovered from their depressive episode (HRS $\leqslant 6$) after 6 weeks of antidepressant treatment did not receive further tricyclic medication for 15 days, after which time a repeat blood sample was taken.

The values obtained for the passive diffusion of tryptophan into the cells at 2 °C were subtracted from the total uptake values obtained at 37 °C to yield the values for the active uptake of tryptophan. The overall uptake, $\bar{y}$, the mean uptake over the five concentrations, was calculated according to the method of *Hallstrom et al.* (1976). The double-reciprocal plot of *Lineweaver* and *Burk* (1934) was used to determine the values of K_m and V_{max}.

The results of the investigation are presented in Table 1. The platelets of the acutely depressed, drug-free patients accumulated significantly more tryptophan than the control group (see $\bar{y}$, Table 1).

Table 1. *Tryptophan uptake characteristics* (K_m, V_{max} *and* $\bar{y}$) *of platelets from controls, depressed patients and recovered depressed patients. Results presented as means ± S.E.M.*

	n	Age	K_m (μM)	V_{max} (pmol/ 10^8 cells/min)	$\bar{y}$ (pmol/ 10^8 cells/min)
Controls	22	47.8 ± 2.7	5.0 ± 0.5	40.7 ± 4.7	11.2 ± 1.0
Depressed patients	20	50.9 ± 2.8	7.7 ± 1.3	54.3 ± 6.5	14.9* ± 1.5
Recovered depressed patients	11	50.5 ± 3.3	4.9 ± 0.7	45.7 ± 10.3	12.7 ± 2.4

* Significantly higher than control subjects, $p < 0.05$.

The data of *Coppen et al.* (1974) suggests that intravenously injected tryptophan disappears more slowly from the circulation of depressed patients than it does in control subjects. The reduced uptake of tryptophan by the tissues of depressed patients has also been demonstrated after debinding experiments with clofibrate (*Wood* and *Coppen*, 1978). Similarily, *Shaw et al.* (1975), in a compartmental analysis of tryptophan metabolism have suggested that the total mass and turnover of tryptophan is reduced in patients who have been or who are suffering from a unipolar affective disorder. The finding of increased accumulation of tryptophan by the blood platelet of depressed patients does not seem to support previous data concerning tryptophan transport in depression. Nevertheless, *Boullin* and *Green* (1972) felt that in general the platelet was not a suitable model for studying the uptake of amino acids by neurones. The ability of the platelet to accumulate tryptophan may be important in the regulation of the plasma levels of this amino acid.

References

Boullin, D. J., Green, A. R.: Mechanisms by which human blood platelets accumulate glycine, GABA and amino acid precursors of putative neurotransmitters. Br. J. Pharmac. *45*, 83—94 (1972).

Coppen, A., Brooksbank, B. W. L., Eccleston, E., Peet, M., White, S. G.: Tryptophan metabolism in depressive illness. Psychol. Med. *4*, 164—173 (1974).

Coppen, A., Swade, C., Wood, K.: Platelet 5-hydroxytryptamine accumulation in depressive illness. Clin. Chim. Acta *87*, 165—168 (1978).

Hallstrom, C. O. S., Rees, W. L., Pare, C. M. B., Trenchard, A., Turner, P.: Platelet uptake of 5-hydroxytryptamine and dopamine in depression. Postgrad. Med. J. *52*, 40—44 (1976).

Hamilton, M.: A rating scale for depression. J. Neurol. Neurosurg. Psychiat. *23*, 56—62 (1960).

Lineweaver, H., Burk, D.: The determination of enzyme dissociation constants. J. Am. Chem. Soc. *56*, 658—666 (1934).

Shaw, D. M., Johnson, A. L., Tidmarsh, S. F., MacSweeney, D. A., Hewland, H. R., Woolcock, N. E.: Multicompartmental analysis of amino acids: 1. Preliminary data on concentrations, fluxes and flow constants of tryptophan in affective illness. Psychol. Med. *5*, 206—213 (1975).

Sneddon, J. M.: Blood platelets as a model for monoamine-containing neurones. Progr. Neurobiol. *1*, 151—198 (1973).

Stahl, S. M.: The human platelet. Arch. gen. Psychiat. *34*, 509—516 (1977).

Wood, K., Coppen, A.: The effect of clofibrate on total and free plasma tryptophan in depressed patients. Neuropharmacol. *17*, 428—430 (1978).

Authors' address: Dr. *K. Wood*, MRC Neuropsychiatry Laboratory, West Park Hospital, Epsom, KT19 8PB, Surrey, U.K.

Journal of Neural Transmission, Suppl. 15, 165—176 (1979)

Distribution of Tryptophan in Erythrocytes, Leukocytes and Thrombocytes, and Its Binding to Plasma Albumin

P. Baumann, M. Perey, S. Laurian, F. Grasset, A. Steck, and **J. M. Gaillard**

Centre de Recherches Psychopathologiques, Clinique Psychiatrique Universitaire de Lausanne, Prilly-Lausanne, Switzerland

With 6 Figures

Summary

The extended theory about a dysfunction of the serotoninergic system in depression and schizophrenia includes the hypothesis of a disturbance in the transport systems of tryptophan and tyrosine from blood to brain. It would be interesting to know if blood cells may be used as a model for the central transport mechanisms of these amino acids. After an oral load, the in vivo distribution of L-tryptophan (50 mg/kg) was studied in the blood plasma, in the different blood cells and its binding to plasma albumin, in six healthy, seven schizophrenic and two depressive subjects.

In all the compartments studied, tryptophan reached a peak, 1—2 hours after the load. Before and after the load, the variation of the tryptophan concentration in the erythrocytes was parallel to the plasma free tryptophan, whereas the uptake of this amino acid was higher in leukocytes and thrombocytes than in erythrocytes.

However, this model does not show differences between schizophrenic and normal subjects with regard to the transport of tryptophan and tyrosine in these cells.

Introduction

In depressive and schizophrenic patients, a disturbance in central serotoninergic structures has been proposed (*Tissot*, 1975; *Murphy et al.*, 1978). In order to understand the factors regulating the metabolism of serotonin in the brain, research has been focussed on the transport mechanisms of its precursor tryptophan from blood to brain. Several authors in this volume give evidence about some essential factors regulating the availability of tryptophan to the brain: on one hand, competition of the other large neutral amino acids with tryptophan at the blood-brain barrier, on the other hand, the reduced availability of tryptophan to the brain by its binding to plasma albumin.

At present, it is not known if blood cells may serve as a peripheral model for transport systems occuring at the blood-brain barrier and in the central nervous system. Comparative studies of the uptake of the amino acids, especially tryptophan by the different blood cells are lacking. Furthermore, it would be interesting to know if transport mechanisms of tryptophan into the erythrocytes, leukocytes and thrombocytes are disturbed in the psychiatric diseases mentioned above.

In this study, the in vivo distribution of tryptophan in the erythrocytes, thrombocytes and leukocytes and its binding to plasma albumin was examined in normal, schizophrenic and depressive subjects, before and after a load of tryptophan or placebo. In parallel, a neurophysiological study was performed, including measurements of the EEG and evoked responses, to test indirectly the uptake of tryptophan in the brain in these experimental conditions. The results of the latter study are presented elsewhere in this volume.

Material and Methods

Six healthy and drug-free volunteers, seven schizophrenic and two depressive patients participated in the biochemical part of this study. While some of the psychiatric patients had been without drugs for at least two weeks, medication was stopped in the other patients, two to seven days before the beginning of the experiment. On the days when experiments were performed, the subjects went without breakfast. On day one, 20 ml EDTA-blood samples were drawn in the morning at 8 a.m. and 11.15 a.m. At 8.30 a.m. the patients ingested 200 ml of milk containing a teaspoon of flour (placebo). On days two and three, the subjects received either placebo or tryptophan for the EEG study. On day four, 20 ml EDTA-blood samples were again obtained at 8 a.m., 9.30 a.m., 10.30 a.m. and 11.30 a.m. At 8.30 a.m., the subjects ingested a suspension of L-tryptophan (50 mg/kg) in 200 ml of milk containing a teaspoon of flour.

Biochemical Determinations

10 ml of the 20 ml blood samples were used for the determination of the tryptophan in the leukocytes, thrombocytes and erythrocytes, where tyrosine was also estimated. The blood cells were separated by the slightly

modified method of *Medzihradski et al.* (1974). Briefly, the erythrocytes were eliminated from the other blood cells by sedimentation at 37 °C after adding plasmagel (Rhodiapharm, Geneva) to the samples. The erythrocytes were further purified in several steps including washings, sedimentation and centrifugation. The thrombocytes were separated from the leukocytes by centrifugation at 80 g and 4 °C, and washed once. The remaining erythrocytes in the fraction of the leukocytes were eliminated by hydrolysis. After washing, the final leukocyte-pellet was virtually free of erythrocytes, but still contained thrombocytes in a ratio of about 0.5 to 2 thrombocytes per one leukocyte. The purity of the fractions was checked by manual counting.

For the isolation of tryptophan in the leukocytes and thrombocytes, 50 μl of the cells suspended in medium were disrupted by sonification and extracted with 1 ml of methanol. The extract was gently evaporated to dryness (Vortex-Evaporator) after centrifugation. The dry residue was dissolved in 1 ml of water and submitted to fluorometric analysis. 500 μl of erythrocytes were extracted by the method of *Curzon* and *Knott* (1975). The final acquous phase containing HCl, the extracts of the thrombocytes and leukocytes as well as total tryptophan were determined fluorometrically (*Flentge et al.*, 1974). The analysis of free tryptophan has been described elsewhere (*Baumann* and *Perey*, 1977). For the determination of tyrosine in the erythrocytes, an aliquot of the washed blood cells was spotted on filter paper and dried. After extraction by water, these samples, and also plasma tyrosine were analysed fluorometrically (*Hochella*, 1967). Proteins (*Lowry et al.*, 1951) and NEFA (*Duncombe*, 1964) were assayed colorimetrically.

Results

In order to find out differences between the groups in the uptake of tryptophan by the blood cells, the levels of total tryptophan were first of all measured in the plasma of our subjects. The normals, schizophrenics and depressives showed no differences regarding this parameter (Fig. 1) either before or after the load of tryptophan. The depressives behaved no differently from the two other groups, for any of the other parameters. For this reason, they are omitted in some of the following figures.

Free tryptophan in the blood plasma and the level of tryptophan in the red blood cells were compared, on the assumption that 1 ml plasma contains 900 mg water.

In the following paragraphs, the ciphers indicated in parentheses are for basic values obtained at 8 a.m. on day one.

There is no difference between normals and schizophrenics, either with regard to the levels of free tryptophan or to the levels of tryptophan in the erythrocytes (Fig. 2). However, if all the subjects

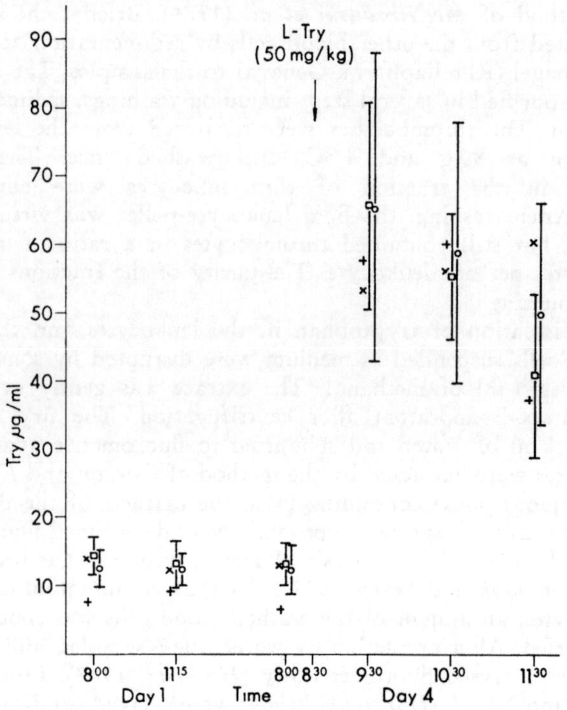

Fig. 1. Total tryptophan levels in normal (□), schizophrenic (○) and depressive subjects (×, +) after placebo (day 1) and 50 mg/kg L-tryptophan (day 4). Mean ± s.d.

are taken as one group, the concentration of this amino acid reaches a higher level ($p < 0.001$) in the erythrocytes (3.95 ± 0.5 µg/900 mg water) than free tryptophan in plasma (2.08 ± 0.54 µg/900 mg water). The levels of tryptophan expressed per 100 mg protein are compared between the erythrocytes (0.79 ± 0.1 µg/100 mg protein), leukocytes (5.74 ± 4.02 µg/100 mg protein) and thrombocytes (6.92 ± 1.8 µg/100 mg protein) (Fig. 3). Before and after the load, the concentration of tryptophan in the red blood cells does not reach the level measured in the leukocytes and thrombocytes. Since one hour after the load, there is a tendency ($p < 0.1$) for tryptophan to be higher in the leukocytes of schizophrenics than in normals, the levels were considered separately:

Fig. 3. Tryptophan levels in leukocytes (----), thrombocytes (——) and erythrocytes (—○—, —□—) of normal (□) and schizophrenic (○) subjects (*cf.* Fig. 1)

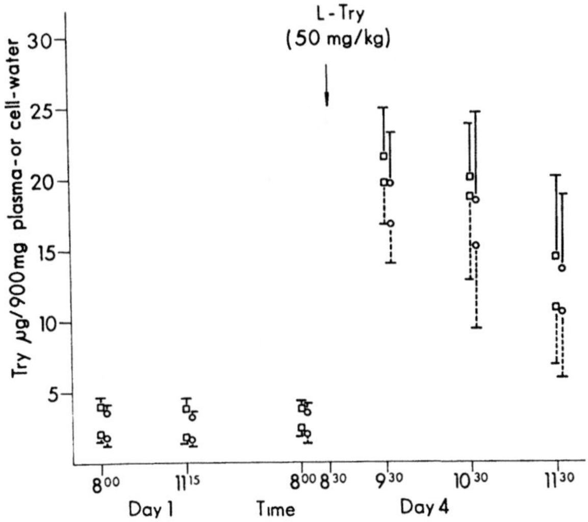

Fig. 2. Free tryptophan (----) and tryptophan levels in the erythrocytes (——) of normal (□) and schizophrenic (○) subjects (*cf.* Fig. 1). The results are expressed per 900 mg water, which is equivalent to 1 ml plasma for free tryptophan

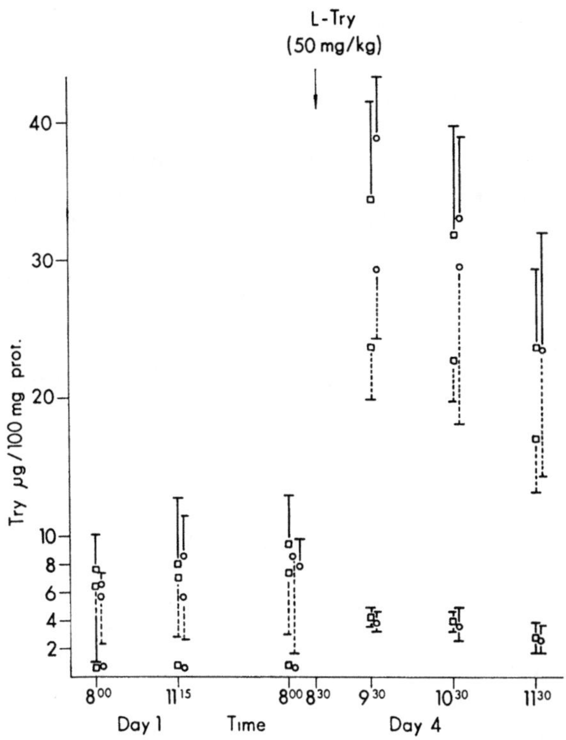

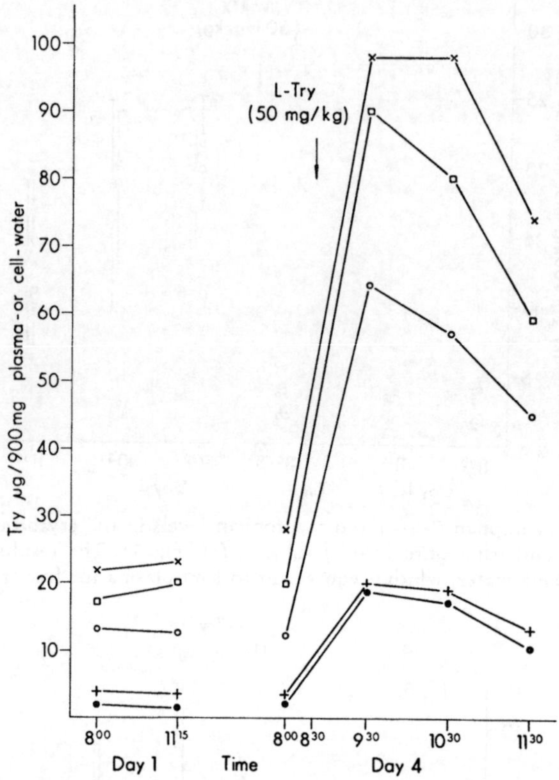

Fig. 4. Mean free (●) and total (○) tryptophan and tryptophan levels in the erythrocytes (+), leukocytes (□) and thrombocytes (×) of all 15 subjects before and after an oral load of L-tryptophan (50 mg/kg). The results are expressed per 900 mg water, which, for free and total tryptophan, is equivalent to 1 ml plasma

1. One hour after the load, the level of tryptophan is higher in the thrombocytes than in the leukocytes, a situation encountered in both groups of subjects ($p < 0.02$ for the normals; $p < 0.01$ for the schizophrenics).

2. This result is found again two ($p < 0.05$) and three hours ($p < 0.1$) after the load, but only in the normal subjects.

In Fig. 3, the results are expressed per 100 mg protein. It would be interesting to compare directly the level of tryptophan in the leukocytes, thrombocytes and erythrocytes. However, the protein content of a cell is not necessarily in a direct relation to the volume of a cell. Thus, the concentration of tryptophan will be expressed per 900 mg cell water (*cf.* Fig. 2). The water content of the blood cells was not determined experimentally, but was derived from values in literature:

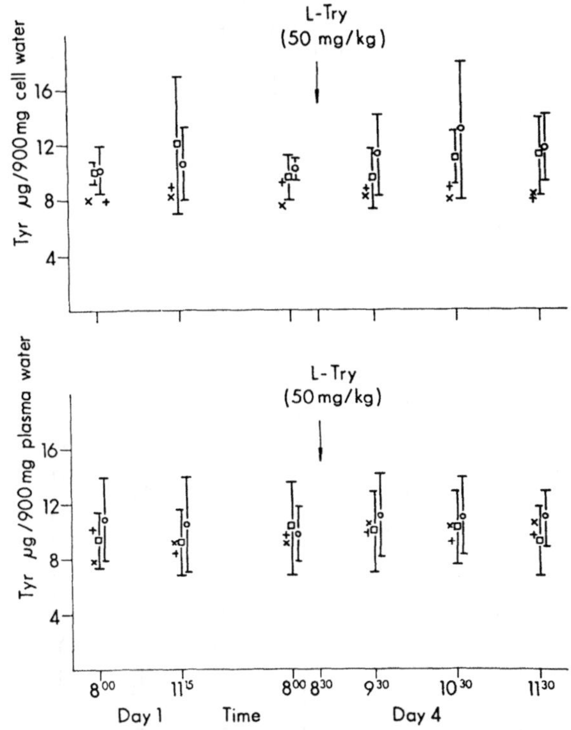

Fig. 5. Tyrosine levels in the erythrocytes (upper graph) and in plasma (lower graph) of normal (□), schizophrenic (○) and depressive (×, +) subjects. The results are expressed per 900 mg water which is equivalent to 1 ml plasma for plasma tyrosine

leukocytes, 64 % water; thrombocytes, 75 % water (*Baron* and *Ahmed*, 1969).

A mean cellular volume of 370 μm^3 and 15 μm^3 and a mean protein concentration of 100 mg/10^9 cells and 3 mg/10^9 cells (personal results) has been taken for leukocytes and thrombocytes respectively. As some uncertainty exists regarding these reference values, only the mean values are represented in Fig. 4, for all subjects taken together. From these semiquantitative data, it may be seen that:

1. Before and after the load, the concentration of tryptophan in the erythrocytes, follows that measured as free tryptophan.

2. Before and after the load, the concentration of tryptophan in the leukocytes and thrombocytes is comparable to the level of total tryptophan.

Tyrosine was also assayed in plasma (10.03 ± 2.52 µg/900 mg water) and in the erythrocytes (9.77 ± 1.42 µg/900 mg water) (Fig. 5).

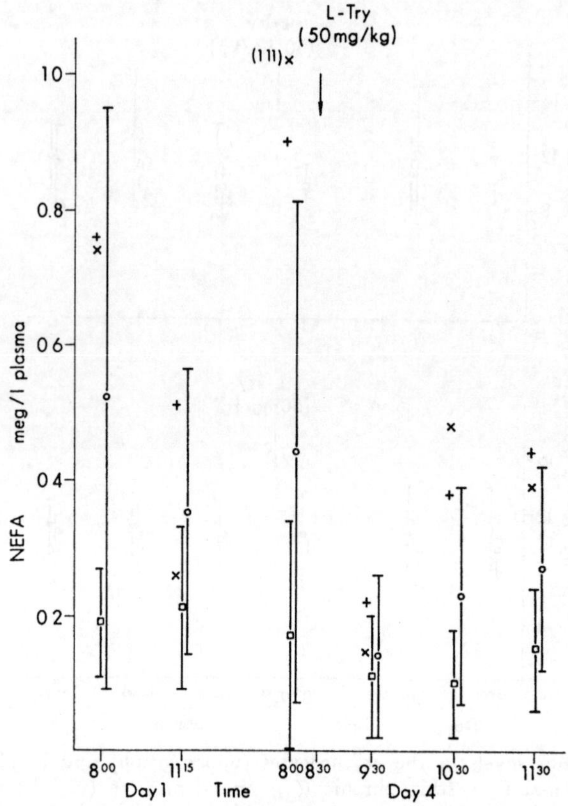

Fig. 6. NEFA-levels in normal (□), schizophrenic (○) and depressive (×, +) subjects

There is no difference between the levels of tyrosine in plasma and in the erythrocytes, before and after the load. Tryptophan itself exerts no effect on tyrosine in the red blood cells or in the plasma.

As the NEFA compete with tryptophan for the binding sites in plasma albumin, their concentration was measured (Fig. 6). There are noteworthy and considerable interindividual differences in the schizophrenic patients and high values in the depressives. These patients may react differently to the stresses inherent in the conditions of the experiment.

Discussion

After ingestion of tryptophan, free and total tryptophan showed a peak during the second hour, confirming the findings of *Greenwood et al.* (1975) and *Domino* and *Krause* (1974). Like the latter authors, we found no differences in the tolerance curves between normals and schizophrenics.

The distribution of tryptophan was studied in the blood cells after separation of the erythrocytes, leukocytes and thrombocytes by the procedure of *Medzihradsky et al.* (1974). This was chosen instead of other possible procedures (*McMenamy et al.*, 1960; *Böyum*, 1968; *Lee et al.*, 1969; *Baron* and *Ahmed*, 1969; *Hsia*, 1972; *Heyneman et al.*, 1975), because it permits recovery of the three blood cell types in a relatively pure state and from a small blood sample. The method of *McEwen et al.* (1968) would probably be the method of choice, but it needs sophisticated equipment. He proposed the separation of the cells by centrifugation with a special rotor, the elutriator.

The procedure used for the present experiments has the disadvantage that the isolation of the cells requires two hours of manipulation. During this time, as the transport processes are rapid, tryptophan and tyrosine may flow out of the cells during the washings. Nevertheless after the load of tryptophan, the accumulation of this amino acid within the blood cells is evident (Figs. 2 and 3). This accumulation may not only depend on transport systems at the cellular membrane, but also on intracellular binding. Indeed, it has recently been shown, that albumin may be synthesized in cultured human lymphocytes (*Goussault et al.*, 1976).

Few papers have dealt with the tryptophan content of the blood cells. In comparison to our baseline levels of tyrosine (60.0 μmol/kg water) and of tryptophan (21.5 μmol/kg water), *McMenamy* (1960) found 72 μmol tyrosine/kg water and 21 μmol tryptophan/kg in the erythrocytes, by paper chromatography and visual examination of the spots on the chromatogram. In the leukocytes, the concentration of tryptophan was estimated to be 222 μmol/kg water by the same author, in contrast to the 118 μmol/kg water that we found. After an incubation time of one hour, the distribution ratio, erythrocytes : plasma was found to be 0.65 : 1 with radioactive D, L-tryptophan (*Winter* and *Christensen*, 1964), in comparison to 0.30 : 1 (present results) with L-tryptophan.

The levels of tryptophan in the erythrocytes differ clearly from those of tyrosine. It is probable, that both compounds use the same transport system. However, tryptophan is bound to the circulating albumin and its affinity is lower for the erythrocytes than for the albumin. This is in contrast to the transport of tryptophan through the blood-brain barrier, where substantial amounts of tryptophan are stripped off the circulating albumin.

The concentration of tryptophan is higher in the leukocytes and thrombocytes than in the erythrocytes. This is in agreement with the lower K_m found for the thrombocytes (*Wood*, this volume) in comparison to the erythrocytes (*Young* and *Ellory* respectively *Rosenberg*, this volume). It is to be noted that only a few studies have been reported in literature concerning the transport of amino acids in leukocytes (*Antonioli et al.*, 1967; *Helmreich* and *Kipnis*, 1962; *Yunis et al.*, 1963; *Rosenberg* and *Downing*, 1965), but none about tryptophan.

In this in vivo experiment, no clearcut differences appeared between schizophrenics and normals with regard to the distribution of tryptophan in the blood plasma and cells. We observed a slight tendency for tryptophan to accumulate more in the leukocytes of schizophrenics than in normals. However, no definite conclusion can be drawn, as the leukocytes were not separated into granulocytes and lymphocytes, whose transport mechanisms may differ and whose distribution may vary from one subject to another (*Yunis et al.*, 1963).

Previously, high NEFA-levels have been reported in depressives (*Müller et al.*, 1970). In the present study, this was also found for the two depressives and some of the schizophrenic patients (Fig. 5). The different reactions to the experimental conditions between normals and patients, may be important for the electrophysiological part of this study (*Laurian et al.*, this volume), where the pharmacological influence of tryptophan may be masked by non-specific effects.

In conclusion, tryptophan is taken up by the circulating blood cells after an oral load. However, this model does not show differences between schizophrenic and normal subjects with regard to the transport of tryptophan and tyrosine in these cells. On the other hand, the molecular basis of the transport systems of the amino acids in the blood cells, especially in the leukocytes, are not yet elucidated. Also, the function of the amino acids in the blood cells remains unclear: are they used only for the synthesis of proteins, or do the blood cells operate as carriers for the amino acids between different organs (*Felig et al.*, 1973)?

References

Antonioli, J. A., Robinson, J. W. L., Vannotti, A.: Amino-acid transport by various leucocyte preparations: A comparative survey. Enzym. biol. clin. *8*, 113—134 (1967).

Baron, D. N., Ahmed, S. A.: Intracellular concentrations of water and of the principal electrolytes determined by analysis of isolated human leucocytes. Clin. Sci. *37*, 205—219 (1969).

Baumann, P., Perey, M.: The analysis of free tryptophan in human blood with the ultrafiltrator: a comparison with other methods. Clin. Chim. Acta *76*, 223—231 (1977).

Böyum, A.: Isolation of mononuclear cells by one centrifugation and of granulocytes by combining centrifugation and sedimentation at 1 g. Scand. J. Clin. Laborat. Invest. *21*, suppl. 97, 77—88 (1968).

Curzon, G., Knott, P. J.: Rapid effects of environmental disturbance on rat plasma unesterified fatty acid and tryptophan concentrations and their prevention by antilipolytic drugs. Br. J. Pharmac. *54*, 389—396 (1975).

Domino, E. F., Krause, R. R.: Free and bound serum tryptophan in drug-free normal controls and chronic schizophrenic patients. Biol. Psychiat. *8*, 265—279 (1974).

Duncombe, W. G.: The colorimetric micro-determination of non-esterified fatty acids in plasma. Clin. Chim. Acta *9*, 122—125 (1964).

Felig, P., Wahren, J., Räf, L.: Evidence of inter-organ amino-acid transport by blood cells in humans. Proc. Nat. Acad. Sci. U.S.A. *70*, 1775—1779 (1973).

Flentge, F., Venema, K., Korf, J.: Automated assay of tryptophan at the nanogram level: determination of tryptophan in cerebrospinal fluid and of total and nonprotein bound tryptophan in serum. Biochem. Med. *11*, 234—241 (1974).

Goussault, Y., Sharif, A., Bourrillon, R.: Serum albumin biosynthesis and secretion by resting and lectin stimulated human lymphocytes. Biochem. Biophysic. Res. Commun. *73*, 1030—1035 (1976).

Greenwood, M. H., Lader, M. H., Kantameneni, B. D., Curzon, G.: The acute effects of oral (—)-tryptophan in human subjects. Br. J. clin. Pharmac. *2*, 165—172 (1975).

Helmreich, E., Kipnis, D. M.: Amino acid transport in lymph node cells. J. Biol. Chem. *237*, 2582—2589 (1962).

Heyneman, R. A., Monbaliu-Bauwens, D., Vercauteren, R. E.: Hydrolytic enzymes in neutrophil and eosinophil leukocytes. Comp. Biochem. Physiol. *50B*, 463—469 (1975).

Hochella, N. J.: Automated fluorometric determination of tyrosine in blood. Anal. Biochem. *21*, 227—234 (1967).

Hsia, D. Y. Y.: Utilization of leukocytes for the study of inborn errors of metabolism. Enzyme *13*, 161—168 (1972).

Lee, M. B., Bolger, C. D., Bridges, J. M.: Distribution of amino-acids in urine, plasma, leucocytes and erythrocytes of leukaemic and normal subjects. Acta Haemat. *42*, 86—93 (1969).

Lowry, O. H., Rosebrough, N. J., Farr, A. L., Randall, R. J.: Protein measurement with the folin phenol reagent. J. Biol. Chem. *193*, 265 to 275 (1951).

McEwen, C. R., Stallard, R. W., Juhos, E. T.: Separation of biological particles by centrifugal elutriation. Anal. Biochem. *23*, 369—377 (1968).

McMenamy, R. H., Lund, C. G., Neville, G. J., Wallach, D. F. H.: Studies of unbound amino acid distributions in plasma, erythrocytes, leukocytes and urine of normal human subjects. J. Clin. Invest. *39*, 1675—1687 (1960).

Medzihradsky, F., Marks, M. J., Metcalfe, J. I.: A simple procedure for the separation of viable blood cells, suitable for long-term "in-vitro" experiments. Biochem. Med. *10*, 153—166 (1974).

Mueller, P. S., Davis, J. M., Bunney, W. E., Weil-Malherbe, H., Cardon, P. V.: Plasma free fatty acids concentration in depressive illness. Arch. Gen. Psychiat. *22*, 216—221 (1970).

Murphy, D. L., Campbell, I. C., Costa, J. L.: The brain serotonergic system in the affective disorders. Prog. Neuro-Psychopharmac. *2*, 1—31 (1978).

Rauen, H.: Biochemisches Taschenbuch, Vol. 2, 2nd ed., p. 343. Berlin-Göttingen-Heidelberg-New York: Springer. 1964.

Rosenberg, L. E., Downing, S.: Transport of neutral and dibasic amino acids by human leukocytes: Absence of defect in cystinuria. J. Clin. Invest. *44*, 1382—1393 (1965).

Tissot, R.: The common pathophysiology of monoaminergic psychoses: A new hypothesis. Neuropsychobiol. *1*, 243—260 (1975).

Winter, C. G.: Christensen, H. N.: Migration of amino acids across the membrane of the human erythrocyte. J. Biol. Chem. *239*, 872—878 (1964).

Yunis, A. A., Arimura, G. K., Kipnis, D. M.: Amino acid transport in blood cells. I. Effect of cations on amino acid transport in human leukocytes. J. Lab. Clin. Med. *62*, 465—476 (1963).

Authors' address: Dr. *P. Baumann*, Centre de Recherches Psychopathologiques, Clinique Psychiatrique Universitaire de Lausanne, CH-1008 Prilly-Lausanne, Switzerland.

Journal of Neural Transmission, Suppl. 15, 177—188 (1979)

The Influence of Oral Tryptophan on Cortical Evoked Responses in Normals and Schizophrenics

S. Laurian, F. Grasset, A. Steck, P. Baumann, and **J.-M. Gaillard**

Clinique Psychiatrique Universitaire de Lausanne, Prilly, Switzerland

With 4 Figures

Summary

The cortical evoked responses to sensory stimulation were studied in order to evaluate the central effects of oral tryptophan in normals and schizophrenics.

In the first study, seven volunteer male students were tested during four experimental sessions. Visual evoked responses to flashes were recorded at occiput and vertex, before and 1 1/2 hours after the ingestion of 35 mg/kg tryptophan in apple purée with 50 g glucose or placebo.

Some significant modifications of evoked responses were observed after tryptophan ingestion: a decrease of amplitude and an increase of variability. The latencies were rather stable, presenting only minor modifications.

From these results, corroborated by some data from literature, it may be supposed that tryptophan induces a decrease of arousal.

In the second study, seven chronic schizophrenics and 6 volunteer controls were tested, using an analogous experimental design, however with some modifications: the tryptophan dose was increased to 50 mg/kg and administered in 200 ml of milk, instead of apple purée with glucose.

The modifications of evoked responses recorded in the preceding study were no longer found. Probably, the substitution of milk for glucose resulted in an alteration of the transport of tryptophan across the blood-brain barrier.

The blood parameters were rather similar in the two studies; therefore, the central effects of tryptophan did not parallel its blood levels.

Only a few studies concerning the effects of tryptophan on cortical evoked responses in man have been performed. *Greenwood et al.* (1975) found that responses to clicks underwent some inconsistent modifications after oral administration of 5 g tryptophan: the latency of P_1 component tended to be reduced, as well as the amplitude of P_2–N_2.

On the other hand, *Hartman* (1977) and *Griffiths et al.* (1972) showed the effectiveness of L-tryptophan in reducing sleep latency. They also observed that this amino acid increased subjective sleepiness in normal subjects at low doses (1—5 g), without producing distortions of physiological sleep.

The purposes of our research were: (a) to study the central effects of tryptophan; (b) to evaluate the biological activity of tryptophan in relation to its level in plasma and in blood cells (*Baumann et al.*, this volume) and (c) to test the hypothesis of a hypoactivity of brain serotoninergic systems in schizophrenia (*Gaillard* and *Tissot*, this volume).

We chose as a model for the study of the central effects the cortical evoked responses to sensory stimulation. Two main kinds of electrical brain activity can be recorded from the surface of the human head: (a) spontaneous activity (EEG) and (b) evoked responses to sensory stimulation. The EEG can be displayed directly after amplification, but the evoked responses must be extracted from the EEG, usually by averaging. The amplitude of evoked responses represents only $^1/_5$ to $^1/_{10}$, or less, of the amplitude of EEG; therefore, single evoked responses are not visible in the background activity. If the evoked response is considered as a signal, the EEG may be considered as a noise. The aim of averaging is to improve the signal to noise ratio. In that way it is possible to obtain average evoked responses (AERs) to different types of stimuli: light, sound, electric shock and others.

With that end in view, we performed two studies in normal subjects and schizophrenics patients.

First Study

Method

Seven volunteer male students were tested, being paid for their participation. Recording of the EEG was performed in an electrically shielded, thermostabilized and partially soundproofed room. Two monopolar leads were used: right occipital and vertex (O_2 and Cz in the 10—20 system) referenced to the interconnected ear lobes. EEG amplification with $^1/_2$ amplitude cut off at 1 Hz and 100 Hz was provided by a Grass Model 78 EEG-polygraph. The data were stored on a PCM system magnetic tape. 100 stimuli at 1 cps generated by a Grass PS2 photostimulator were displayed on a translucent screen placed 30 cm in front of the subject's eyes. The stored data were filtered (active filters, cut off at 40 Hz) and processed using a Nicolet Med-80 minicomputer. The EEG of 500 ms following the onset of the stimulus was sampled at 250 Hz. Individual averages of 100 responses each, as well as the general average of all responses of all subjects ("grand mean") were calculated for each experimental condition and lead. The latencies and amplitudes of evoked response components were measured using a cursor routine of the minicomputer. Four components labelled N_1, P_1, N_2 and P_2 were identified at O_2, whereas three components labelled P_1, N_1 and P_2 were identified at Cz (Fig. 1). The amplitude and the variability (Amp/SD and SD/trials, calculated after *Callaway* [1973]) of individual averages were evaluated between 50—150 ms and 150—250 ms for the occipital lead and between 70—220 ms for the vertex lead.

Experimental Design

Each subject was submitted to four experimental sessions:

1st session—accomodation to experimental conditions in EEG laboratory;

2nd session—biochemical investigations;

3rd session—electrophysiological investigations;

4th session—electrophysiological investigations.

The last two sessions consisted of a basal recording of brain electrical activity, followed by the ingestion of tryptophan (35 mg/kg in apple purée with 50 g glucose) or placebo (apple purée with 50 g glucose) and $1\,^1/_2$ hours later, by a second recording of the brain electrical activity. Significance of differences was established using a t-test for related samples (before versus after tryptophan or placebo).

Results

a) Grand Means (Fig. 1 and Table 1)

Each curve represents the average of 700 responses. Latencies were rather stable, presenting only minor modifications; a slight increase of N_2 and P_2 latencies at O_2 was observed after tryptophan. By contrast, the amplitudes showed a marked decrease after tryptophan for the two leads. The N_1–P_1 and P_1–N_2 components at O_2 were decreased from 7.5 to 3.7 μV and from 5.3 to 2.2 μV respectively. Similar diminutions were recorded at vertex for the components P_1–N_1

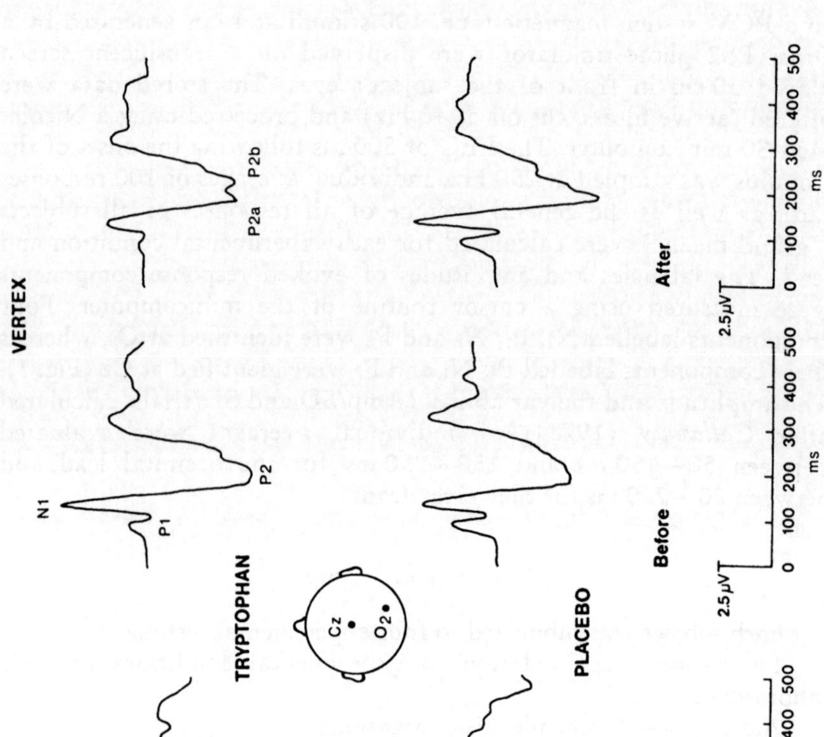

Fig. 1. Visual evoked responses (grand means) before and after tryptophan ingestion (35 mg/kg in apple purée with 50 g glucose). Each curve represents the average of all responses of all subjects; the amplitudes of different components decrease after tryptophan

(5.5 μV before, 2.7 μV after tryptophan) and N_1–P_2 (10.9 μV before, 7.6 μV after tryptophan). No such modifications were observed after placebo.

Table 1. *The peak-to-peak amplitudes of different components of visual evoked potentials recorded at occiput and vertex before and after tryptophan ingestion (35 mg/kg in apple purée with 50 g glucose). The amplitudes were measured on the grand means obtained from the 7 subjects*

CONDITIONS		AMPLITUDE (μV)				
		Occipital			Vertex	
		$N_1 - P_1$	$P_1 - N_2$	$N_2 - P_2$	$P_1 - N_1$	$N_1 - P_2$
TRYPTOPHAN	Before	7.5	5.3	6.2	5.5	10.9
	After	3.7	2.2	4.6	2.7	7.6
PLACEBO	Before	6.8	4.3	3.5	4.0	8.6
	After	7.8	5.0	4.3	5.1	10.6

b) Individual Averages

In order to evaluate the individual contributions to the grand means, we superimposed all the individual averages for each experimental condition (Fig. 2). It appears that the modifications observed after tryptophan were systemically, being present in most subjects. Moreover, the interindividual variability increased after tryptophan.

The quantitative evaluation of individual averages gave the following results (Table 2):

(1) Amplitude (Amp/SD)—a significative decrease at O_2 between 50—150 ms and at vertex between 70—220 ms was observed after tryptophan.

(2) Intraindividual variability (SD/trials)—a significant increase at Oz between 150—250 ms and at vertex between 70—220 ms was observed after tryptophan. None of the modifications recorded after placebo were significant. So, it appears that oral tryptophan decreases the amplitude and increases the inter- and intraindividual variability of visually cortical evoked responses.

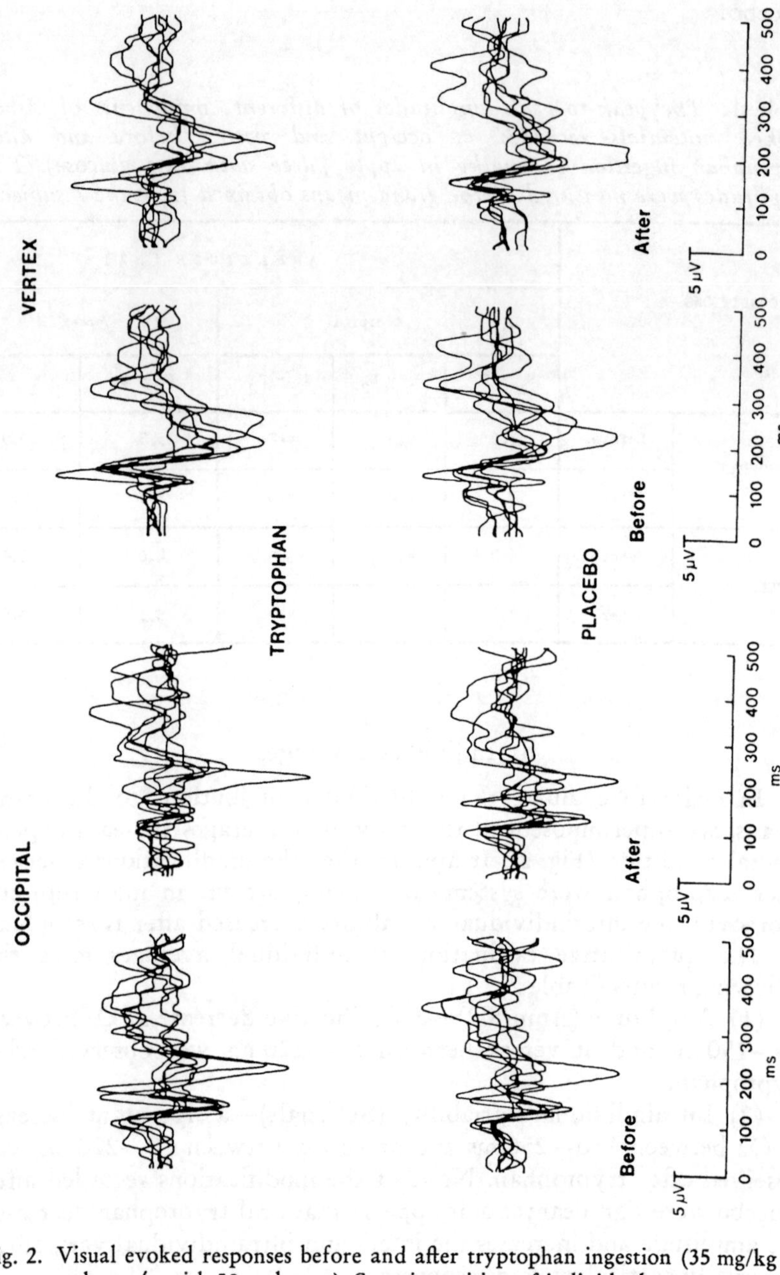

Fig. 2. Visual evoked responses before and after tryptophan ingestion (35 mg/kg in apple purée with 50 g glucose). Superimposition of individual averages

Table 2. *The changes in amplitude and intraindividual variability after tryptophan ingestion (35 mg/kg in apple purée with 50 g glucose). The calculations were performed between the limits displayed at the right side; 7 subjects*

Lead	Limits (ms)	Experimental condition	Amplitude			Variability		
			Before	After	P	Before	After	p
Occipital (O_2)	50 - 150	Tryptophan	25,5	20,4	.05	22,5	24,9	ns
		Placebo	23,7	25,0	ns	23,2	23,4	ns
	150 - 250	Tryptophan	25,3	25,8	ns	20,3	22,8	.05
		Placebo	23,3	23,6	ns	19,1	21,1	ns
Vertex (C_z)	70 - 220	Tryptophan	30,1	26,0	.01	22,1	24,5	.01
		Placebo	28,2	28,4	ns	20,6	22,1	ns

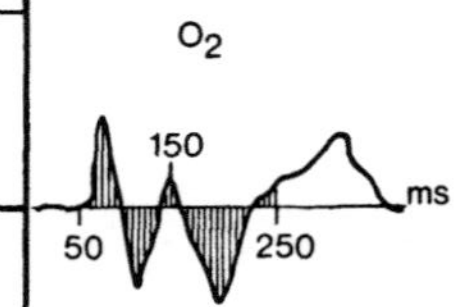

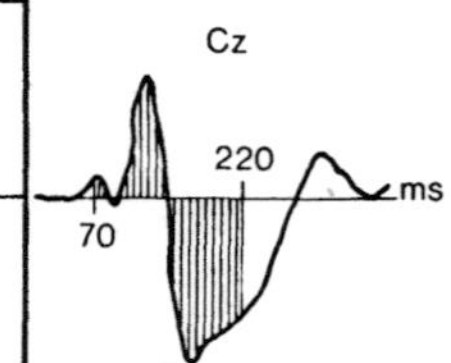

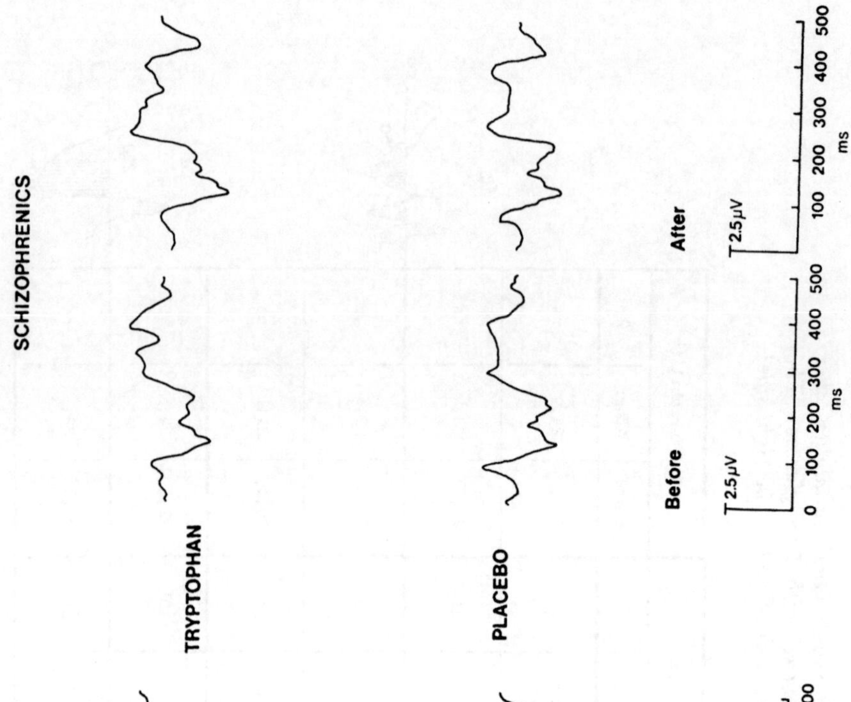

Fig. 3. Visual evoked responses (grand means) recorded at occiput in normals and schizophrenics before and after tryptophan ingestion (50 mg/kg in 200 ml of milk); no significant changes in amplitude are observed. Patients present lower amplitudes than normals

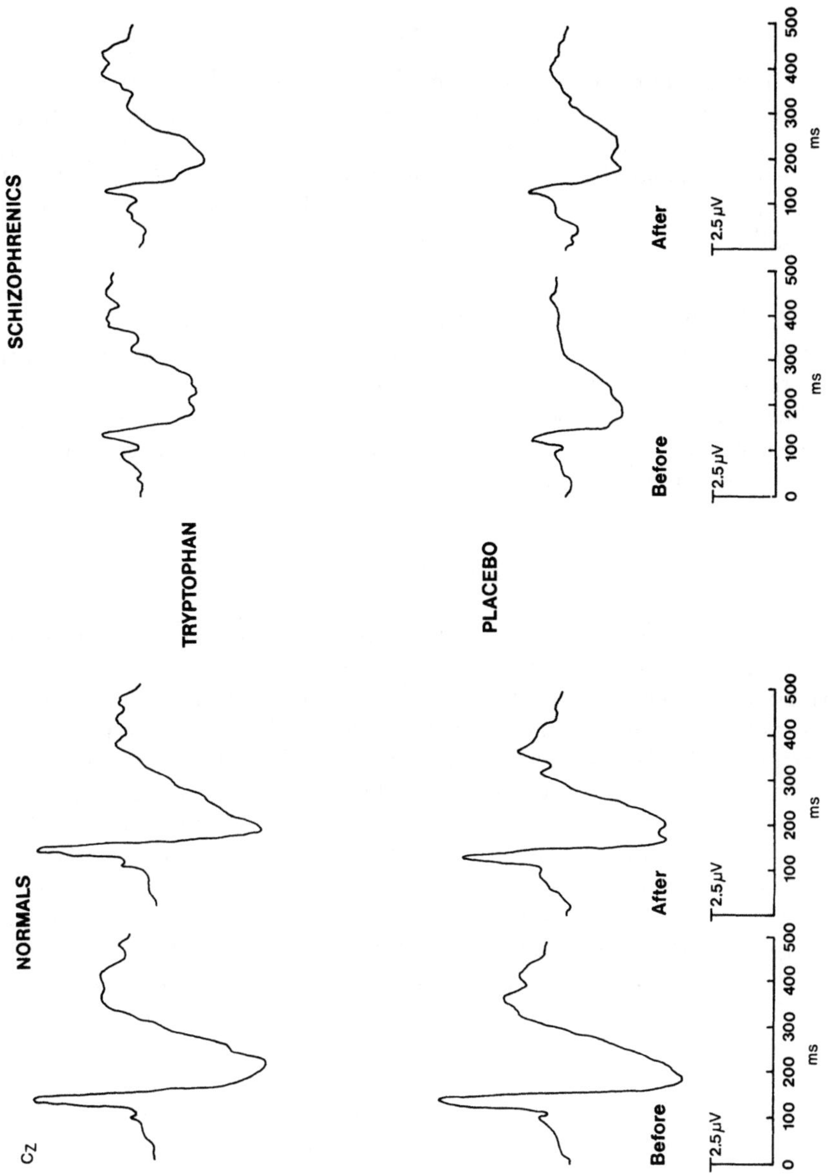

Fig. 4. Visual evoked responses (grand means) recorded at vertex in normals and schizophrenics before and after tryptophan ingestion (50 mg/kg in 200 ml of milk); no significant changes in amplitude are observed. Patients present lower amplitudes than normals

Second Study

Method

It was analogous to that used in the first study; however, some modifications were introduced in the experimental design. A 5th session for biochemical investigations was added and the tryptophan dose was increased to 50 mg/kg and administered in 200 ml of milk (instead of apple purée with glucose); the placebo consisted of 200 ml of milk.

Results

Only the grand means were calculated. The modifications of evoked responses recorded in the preceding study were no longer found either in normals or in schizophrenics (Figs. 3 and 4). On the other hand, it may be observed that the visually evoked cortical responses are modified in schizophrenics. The most conspicuous feature was the low amplitude of evoked responses in these patients.

Discussion

From the first study it may be inferred that oral tryptophan induces some definite modifications of cortical visually evoked responses, namely: (a) a decrease of amplitude and (b) an increase of variability. These modifications were observed both at occipital and vertex leads. It has been suggested that the late components of visually evoked responses present at the occiput and vertex, represent different neurophysiological phenomena (*Lehtonen,* 1973). The occipital deflections probably reflect the function of the modality-specific area and the vertex waves, the function of a modality-non-specific area. Several authors pointed out the differences between occipital and vertex responses (*Kooi* and *Bagchi,* 1964; *Vaughan,* 1969; *Perronet* and *Gerin,* 1973; *Laurian et al.,* 1973; *Kress,* 1976).

It appears that tryptophan influences cortical areas having different functional roles; therefore, it seems to have a rather general action.

Concerning the significance of the observed modifications of evoked responses, it might be supposed that tryptophan induces a decrease of arousal. This opinion corroborates the data presented by *Hartman* (1977), demonstrating that tryptophan reduces sleep latency and increases subjective sleepiness. On the other hand, although the interrelationship between attention, arousal, and evoked responses are complex, it is generally admitted that a decrease of arousal is accompanied by an increase of the variability and a diminution of the amplitude of some components of evoked responses.

A rather surprising finding was the absence of effects of tryptophan on the evoked responses in the second study, whereas in the first one, some definite modifications of evoked responses were recorded after tryptophan ingestion. This inconsistency might be related to the changes introduced in the design of the second study: the dose of tryptophan was increased from 35 mg/kg to 50 mg/kg and tryptophan was incorporated in 200 ml of milk, instead of apple purée with glucose. It is very improbable that the tryptophan dose accounts for this difference. It is more likely due to the substitution of milk for glucose. This modification probably results in an alteration of the transport of tryptophan across the blood-brain barrier.

It must be noted that the blood parameters were rather similar in the two studies (*Baumann et al.*, this volume; *Baumann et al.*, in press), whereas central responses were quite different. Consequently, it appears that the central effects of tryptophan do not parallel its blood levels.

Conclusions

From these studies it may be inferred that oral tryptophan induces some definite modifications of cortical evoked responses to visual stimuli: (a) a decrease of amplitude and (b) an increase of variability. These alterations were recorded at occiput and vertex, for components between 50 and 250 ms. These components appear to be related to association cortical areas.

From our results, corroborated by some data from the literature, it may be supposed that tryptophan induces a decrease of arousal.

On the other hand, the central effects of tryptophan do not parallel its blood level.

No conclusions can be drawn from these studies on the hypothesis of a hypoactivity of brain serotoninergic systems in schizophrenia.

References

Baumann, P., Perey, M., Laurian, S., Grasset, F., Steck, A., Gaillard, J.-M.: Distribution of tryptophan in the erythrocytes, leucocytes and thrombocytes, and its binding to plasma albumin. J. Neural Transm., Suppl. 15, pp. 165—176. Wien-New York: Springer. 1979.

Baumann, P., Perey, M., Laurian, S., Gaillard, J.-M.: Surcharge de tryptophane chez l'homme: Sa distribution dans les cellules sanguines. In: Comptes rendus du 4e colloque intern. de Pont-à-Mousson, 1978 (in press).

Callaway, E., Halliday, R. A.: Evoked potential variability: effects of age, amplitude and methods of measurement. Electroenceph. clin. Neurophysiol. *34*, 125—133 (1973).

Greenwood, M. H., Lader, M. H.: The acute affects of oral (—)-tryptophan in human subjects. Br. J. clin. Pharmacol. *2*, 165—172 (1975).

Griffiths, W. J., Lester, B. K., Coulter, J. S., Williams, H. L.: Tryptophan and sleep in young adults. Psychophysiology *9*, 345—356 (1972).

Hartman, E.: L-tryptophan as an hypnotic agent: a review. Waking and Sleeping *1*, 155—161 (1977).

Kooi, K. A., Bagchi, B. K.: Observations on photically evoked occipital and vertex waves during the sleep in man. Amer. N.Y. Acad. Sci. *112*, 270—280 (1964).

Kress, G.: Effects of visual stimulation on the evoked response recorded at the vertex. Brain Res. Bull. *1*, 337—339 (1976).

Laurian, S., Reigner, J., Heimann, H., Gruber, G., Wertheimer, J., Lobrinus, A.: Visual evoked responses in elderly psychiatric patients. Electroenceph. clin. Neurophysiol. *34*, 760 (1973).

Lehtonen, J. B.: Functional differentiation between late components of visual evoked potentials recorded at occiput and vertex: effect of stimulus interval and contour. Electroenceph. clin. Neurophysiol. *35*, 75—82 (1973).

Peronnet, F., Gerin, P.: Potentiels évoqués auditifs et visuels: topographie et interactions. In: Activités évoquées et leur conditionnement chez l'homme normal et en pathologie mentale, pp. 35—77. Paris: INSERM. 1973.

Vaughan, H. G., jr.: The relationship of brain activity to scalp recordings of event-related potentials. In: Average Evoked Potentials (*Donchin, E., Lindsley, D. B.*, eds.), pp. 45—94. Washington: NASA SP-191. 1969.

Authors' address: Dr. *S. Laurian*, Clinique Psychiatrique Universitaire de Lausanne, CH-1008 Prilly, Switzerland.

Journal of Neural Transmission, Suppl. 15, 189—196 (1979)

Blood-Brain Movements of Tryptophan and Tyrosine in Manic-Depressive Illness and Schizophrenia

J.-M. Gaillard and **R. Tissot**

Clinique Psychiatrique Universitaire de Genève, Chêne-Bourg, Switzerland

Summary

In order to investigate possible disturbances of the blood-brain transport mechanisms of monoamine precursors in manic-depressive illness and schizophrenia, we have measured the brain arterio-venous difference of DOPA, or 5-HTP, or tyrosine and tryptophan in 36 patients, during the infusion of either L-DOPA or L-5-HTP. The infusion lasted for 30 min, and blood was sampled during and immediately after the infusion, simultaneously in the femoral artery, the jugular vein and a vein of the arm. During the infusion of L-DOPA, manic patients have a higher extraction of L-DOPA than depressive patients and controls. During the infusion of L-5-HTP, depressive patients have a higher brain extraction of 5-HTP than manic or schizophrenic patients. In depressive patients, a small uptake of tryptophan correlated with a large outflow of tyrosine was observed. The opposite was seen in manic patients, with an outflow of tryptophan correlated with an uptake of tyrosine. In schizophrenics, there was an outflow of tryptophan and random variations of tyrosine. These brain arterio-venous differences were not correlated with arterio-venous differences for peripheral tissues. Taken together, these results are compatible with a disturbance of the blood-brain transport of amino acids precursors of monoamines in manic-depressive illness and schizophrenia.

Introduction

Many lines of evidence suggest a dysfunction of brain monoaminergic (MA) systems in the physiopathology of two major human psychoses: manic-depressive illness and schizophrenia (for references, see *Tissot*, 1975). Some kind of genetic vulnerability on the one hand, environmental, developmental or psychological disturbances on the other, may both be involved, but their relative contribution to the disease is not known. The demonstration of an abnormal function of MA pathways does not necessarily point to a genetic defect. It can be reasonably supposed that a long lasting unfavourable influence of external origin, even in the absence of any endogenous component, also has consequences on the level of activity of these systems. Another possibility is a particular environmental condition occurring at a critical moment of the development of the individual and inducing a permanent fragility.

The arguments advanced in favor of the MA theories are mostly of a pharmacological nature. Direct measurement of MA metabolism in the brain is unfortunately impossible. Urine and blood are too heavily biased by contributions of the whole body, with the possible exception of methoxy-hydroxyphenylglycol (MHPG), a catabolite of noradrenaline, which is believed to be mainly produced in the brain. The analysis of cerebro-spinal fluid (CSF) has given variable results, possibly because the level of monoamine metabolites in CSF not only results from the activity of brain MA systems, but also from the activity of transport mechanisms responsible for the exchange of these substances between brain, blood and CSF. The study of autopsy material is submitted to the same constraints. In the brain of patients deceased from suicide, a decrease of 5-HT and of 5-HIAA has often been reported (*Bourne et al.*, 1968; *Lloyd et al.*, 1974), but this may be explained either by decreased activity of brain 5-HT systems or, on the contrary, by enhanced activity of the same systems along with enhanced turnover of the amine and enhanced activity of its transport mechanism (*Tissot*, 1975).

A further difficulty originates from the heterogeneity of the concerned pathological conditions. It is conceivable that clinical symptomatology does not accurately reflect biochemical abnormalities, unless there is strong evidence of a causative relation. It follows that the same biochemical disturbance may have several different clinical expressions, depending, for instance, on the participation of environmental factors; conversely, different biochemical disturbances may converge into a more or less common symptomatology. Thus, a clinically-based sorting of the patients can be misleading by mixing different types of biological dysfunctions. In the absence of a more precise knowledge, however, we must rely at present mostly upon clinical classifications.

The activity of brain MA systems can be affected by variations in the availability of their precursors, pointing to a possible involvement of the blood-brain transport mechanism (*Wurtman* and *Fernström*, 1972). Therefore, we have measured the brain arterio-venous differerece (BAVD) of tyrosine, L-DOPA, tryptophan and L-5-HTP in several categories of patients (*Geissbühler et al.*, 1972; *Gaillard et al.*, 1973; *Tissot et al.*, 1978). By BAVD, we understand the difference in the plasma level of a given precursor between the arterial blood and the blood of the jugular vein. A positive BAVD means that the plasma concentration of the precursor decreases during the passage of blood through the brain capillaries (extraction), and in the case of a negative BAVD, the concentration of the precursor is higher in the venous blood leaving the brain (outflow or backflow). Exchanges are taking place between blood and another compartment, the nature of which we cannot specify: endothelial cells or the capillary walls, glia, neurons, extracellular space. Possible exchanges with blood cells must also be considered.

If they exist, the disturbances of blood-brain transport mechanisms are probably small. In order to enlarge them, we have studied the BAVD during the load with a precursor by infusion in a peripheral vein. In most cases, the BAVD was compared to a similar arterio-venous difference in the periphery. In a first series of experiments, we wished to bypass the specific step in monoamine metabolism, namely the hydroxylation; therefore, we infused L-DOPA or L-5-HTP and studied the BAVD of these amino acids. The results encouraged us to study the endogenous precursors tyrosine and tryptophan during load with L-DOPA and L-5-HTP, since it is generally admitted that L-DOPA and L-5-HTP are transported by the same mechanism as are tyrosine and tryptophan (*Lajtha*, 1968).

Methods

The experiments described here were carried out in 36 patients: 5 controls (alcoholics), 10 manic-depressive patients in depressive phase, 10 in manic phase, and 11 schizophrenics. They were aged 25 to 62 years, with approximately the same distribution for all categories. All patients were hospitalized, most of them in our clinic; some of them were studied in the Psychiatric Service of the National Institute of Neurology in Mexico. The diagnoses were established by at least two psychiatrists, on the basis of clinical interviews and of the previous history of the disease (*i.e.* anterior manic or depressive phases for manic-depressive patients). Their previous medication was discontinued and they were given only a small dose of Diazepam for 3—4 days. It was not possible to maintain them completely drug-free because of their generally heavy symptomatology.

On the day of blood sampling, they were lightly anesthesized with pentobarbital, and catheters were placed in the femoral artery, in the jugular vein and, in most cases, in a vein of the arm. After complete recovery from anesthesia, L-DOPA or L-5-HTP were infused i.v. in the other arm with a precision pump during 30 min. In experiments where DOPA or 5-HTP were studied, these precursors were given at the dose of 0.10 mg/kg/min (L-DOPA) and 0.11 mg/kg/min (L-5-HTP). In experiments where tyrosine and tryptophan were assayed, the dose of L-DOPA or L-5-HTP was 0.015 mg/kg/min. Blood samples were withdrawn simultaneously from the artery and from the veins at the following times from onset of infusion: 30 sec; 2 min 30 sec; 5 min; 10 min; 20 min; 30 min, and finally 2 min after the end of infusion. In some cases, blood was continuously obtained between min 4 and 30 from the onset of infusion. In tyrosine/tryptophan experiments, a blood sample was withdrawn before the beginning of the infusion, in order to establish the basal level of these amino acids. This was not necessary for DOPA and 5-HTP, since they are not present in the blood at levels detectable by our techniques. The amino acids were assayed using classical fluorometric methods (*Udenfriend et al.*, 1955; *Wong et al.*, 1964; *Maickel*, 1972; *Williams et al.*, 1972; *Eccleston*, 1973). In order to assess the role of peripheral decarboxylation of the injected precursors, some patients were treated with benserazide, a decarboxylase inhibitor (750 mg, 12 hours before and 500 mg, 45 min before the beginning of the infusion).

Results and Discussion

During the infusion of L-DOPA, the plasma level of this amino acid showed a relatively slow increase, reaching a maximum concentration of about 5 μg/ml. This level was lower in manic than in depressive patients and controls. After the end of the infusion, plasma DOPA quickly fell again to more than 50 % of its maximum value, that is in 2 min. Pretreatment with benserazide almost doubled the level reached during infusion. Manic patients again had a lower level than depressive and controls, but the difference was smaller than without benserazide. During the infusion of L-5-HTP, the plasma level of this amino acid rose much more (to about 15 μg/kg) and this rise was not influenced by pretreatment with benserazide. The fall of plasma level after the end of the infusion was slower than the fall of DOPA. There were no marked variations between the diagnosis categories, but depressive patients reached values among the highest. The differences between L-DOPA and L-5-HTP, as well as their different behaviour after peripheral decarboxylase inhibition probably resulted from the greater affinity of L-DOPA for the enzyme decarboxylase (*Lovenberg et al.*, 1962).

The BAVD was calculated by integration of the arterio-jugular difference during the whole experiment, divided by the duration of the experiment. The values of BAVD are given in Table 1, in ng/ml/min. A puzzling finding was a negative BAVD in the 2 depressive patients and in the control during infusion of L-DOPA without benserazide. On the contrary, the 2 manic patients had a positive BAVD. After benserazide, all patients had a positive BAVD, the lowest in controls, the highest in manic subjects. Thus, the difference between controls, depressive and manic patients remained the same, but was entirely shifted to the positive side.

Table 1. *Brain arterio-venous differences of precursors of monoamines in different categories of patients during an infusion of 30 min of the precursor indicated in the left column. These values have been obtained by integration of the arterio-venous difference during the whole experiment and divided by the duration of the experiment. They express the number of ng of the amino-acid leaving (positive difference) or entering (negative difference) each ml of plasma during each minute of the infusion. The values indicated are averages of the number of patients in brackets. All individual values had the same sign as the average, except in the last series of 8 schizophrenics for the brain artcrio-venous difference of tyrosine; in this last group, 4 patients had an uptake of tyrosine and the other 4 had an outflow*

			Brain arterio-venous difference (ng/ml/min)			
Treatment	Patients	(n)	DOPA	5-HTP	TYR	TRY
L-DOPA	Control	(1)	—265	—	—	—
(0.10 mg/kg/min)	Depression	(2)	— 68	—	—	—
	Mania	(2)	+155	—	—	—
L-DOPA + DCI*	Controls	(2)	+ 82	—	—	—
(0.10 mg/kg/min,	Depression	(3)	+203	—	—	—
1250 mg)	Mania	(3)	+380	—	—	—
L-5-HTP	Control	(1)	—	+369	—	—
(0.11 mg/kg/min)	Depression	(2)	—	+630	—	—
	Mania	(2)	—	+239	—	—
	Schizophrenia	(3)	—	+326	—	—
L-5-HTP + DCI*	Controls	(2)	—	+239	—	—
(0.11 mg/kg/min,						
1250 mg)						
L-5-HTP or	Depression	(3)	—	—	—3111	+196
L-DOPA	Mania	(3)	—	—	+1060	—567
(0.015 mg/kg/min)	Schizophrenia	(8)	—	—	+2489	—567

* DCI: Decarboxylase Inhibitor (benserazide).

The negative BAVD in depressives and in the control is not a technical artifact. The reproducibility of the biochemical assay is good (± 5 %), and their specificity has been verified by chromatography. The two main spots correspond to DOPA and O-methyl-Dopa (which was also assayed in these experiments); the other spots correspond to tyrosine, phenylalanine, and aliphatic amino acids which do not interfere with our fluorescence technique. In addition, we have studied 3 control subjects by injection of L-(2-^{14}C)-DOPA. The BAVD of plasma amino acids radioactivity was positive in the first two minutes and became slightly negative in the following two minutes. The BAVD of total radioactivity however remained positive. This seems to indicate that in the first two minutes, there is an extraction of DOPA. This DOPA is subsequently partly released, giving a negative BAVD. This mechanism, namely a rapid extraction of DOPA during the first few minutes of infusion, with a subsequent backflow, can be responsible for the negative BAVD. This was not observed after peripheral decarboxylase inhibition or after 5-HTP, which points to a possible involvement of brain capillary walls.

During infusion of L-5-HTP all BAVD's were positive. Their numerical values were similar in controls, in manic and in schizophrenic patients, but markedly higher in depressive patients. Pretreatment with benserazide was without effect in the 2 controls studied in this condition.

One of the schizophrenic patients under L-5-HTP had a relatively high positive BAVD (608 ng/ml/min). This value was very close to those obtained in depressive patients. This patient, a woman of 51 years, presented a rather typical aspect of schizophrenia, but a couple of months later she had a melancholic episode for the first time in her life, and was responsive to chlorimipramine.

Thus, during an infusion of L-DOPA, manic patients show a greater extraction of this precursor than the other patients, whereas during an infusion of L-5-HTP, the depressives have a greater extraction of 5-HTP. A possible interpretation of these findings is an enhanced activity of the brain catecholaminergic (CA) systems in mania, linked to a greater activity of the transport mechanisms for CA precursors, and conversely an enhanced 5-HT activity in depression, also linked to a greater activity of the transport mechanisms for the precursors of 5-HT. Therefore, we decided to study the BAVD of the respective endogenous precursors, tyrosine and tryptophan, after a small load with L-DOPA or L-5-HTP. This study was a first attempt to verify the model proposed by one of us (*Tissot*, 1975). This model postulates a brain uptake of tryptophan increased in depression, normal or decreased in mania, and decreased in schizophrenia, and a brain uptake of tyrosine decreased in depression, enhanced in mania and in schizophrenia.

An infusion of L-DOPA influenced the movements of tyrosine and tryptophan in a similar way as did an infusion of L-5-HTP. In depressives, a relatively small uptake of tryptophan with a large outflow of tyrosine was observed. The opposite situation was seen in manic patients, with an outflow of tryptophan and an uptake of tyrosine. In both categories of patients the BAVD of tryptophan and of tyrosine were highly correlated: r = 0.968 (depression), and r = 0.999 (mania). However, these correlations, calculated with 3 cases each, must be viewed with caution. Nevertheless, the difference in schizophrenics was striking: these cases showed a negative BAVD of tryptophan and a randomly distributed BAVD of tyrosine, negative in 4 cases and positive in the remaining 4. There was no correlation between the movements of tryptophan and tyrosine in schizophrenics. Thus, in manic-depressive illness, a movement of tryptophan in one direction is accompanied by a movement of tyrosine in the other. In addition, the movements (outflow) of tyrosine in depression are much greater, with respect to the movements of tryptophan, than in mania. On the contrary, the blood-brain movements of these amino acids are not linked in schizophrenia, a disease in which the only consistent finding is an outflow of tryptophan.

These findings are in favour of the above mentioned hypothesis. They cannot be accounted for by an exchange between plasma and blood cells, because there is no correlation between the BAVD and the peripheral arterio-venous differences in any of the conditions of this study.

Conclusion

Taken together, these results support the view that a disturbance in the blood-brain transport mechanism of precursors of monoamines could play a role in the physiopathology of manic-depressive illness and schizophrenia. However, since the membrane transport of these two amino acids is competitive (*Battistin et al.*, 1971), it is not clear why the relation between them is different in manic-depressive illness and in schizophrenia. Further, these membrane movements are complex and the BAVD is actually the algebraic sum of uptake, outflow and exchange. An attempt to separate these components could throw more light on the physiopathology of these two psychoses.

References

Battistin, L., Grynbaum, A., Lajtha, A.: The uptake of various amino-acids by the mouse brain in vivo. Brain Research *29*, 85—99 (1971).

Bourne, H. R., Bunney, W. E., Colburn, R. W., Davis, J. M., Davis, J. N., Shaw, D. M., Coppen, A. J.: Noradrenaline, 5-hydroxytryptamine, and 5-hydroxyindoleacetic acid in hindbrains of suicidal patients. Lancet *II*, 805—808 (1968).

Eccleston, E. G.: A method for the estimation of free and total acid-soluble plasma tryptophan using an ultrafiltration technique. Clinica chim. Acta *48,* 269—272 (1973).

Gaillard, J.-M., Eisenring, J.-J., Tissot, R.: Consommation cérébrale et périphérique de L-5-HTP chez des patients atteints de syndromes maniaques, dépressifs ou psychotiques. Encéphale *62,* 395—407 (1973).

Geissbuhler, F., Eisenring, J.-J., Friedli, P., Bartholini, G., Tissot, R.: Consommation cérébrale et périphérique de L-DOPA chez des patients atteints de syndromes parkinsoniens, dépressifs ou maniaques sous L-DOPA en perfusion combinée ou non à un inhibiteur de la décarboxylase. Encéphale 2, 127—148 (1972).

Lajtha, A.: Transport as control mechanism of cerebral metabolite levels. In: Progress in Brain Research, Brain-Barrier Systems (*Lajtha, A., Ford, D. H.,* eds.), 29, pp. 201—218. Amsterdam: Elsevier. 1968.

Lloyd, K. G., Farley, I. J., Deck, J. H. N., Hornykiewicz, O.: Serotonin and 5-hydroxyindoleacetic acid in discrete areas of the brainstem of suicide victims and control patients. Adv. Biochem. Psychopharmacol. *11,* 387—397 (1974).

Lovenberg, W., Weissbach, H., Udenfriend, S.: Aromatic L-amino acid decarboxylase. J. biol. Chem. *237,* 89—93 (1962).

Maickel, R. P.: Fluorometric analysis of 5-hydroxytryptamine and related compounds. Methods Neurochem. *2,* 101—129 (1972).

Tissot, R.: Un concept hypothétique: l'unité physiopathologique des psychoses monoaminergiques. Encéphale *I,* 289—339 (1975).

Tissot, R., Castellanos, G., Gaillard, J.-M., Estrada, E., Eisenring, J.-J., Boleaga, B., Hyde, A., Miranda, R., Hovaguimian, T.: Capture du tryptophane et de la tyrosine plasmatiques dans quelques cas de psychose maniaco-dépressive et de schizophrénie. Neuropsychobiology *4,* 65—73 (1978).

Udenfriend, S., Titus, K., Weissbach, H.: The identification of 5-HIAA in normal urine and a method for its assay. J. Biol. chem. *216,* 499—505 (1955).

Williams, C. H., King, D. J., Cairns, J.: Microestimation of tryptophan in plasma by a fluorometric procedure. Biochem. Med. *6,* 504—507 (1972).

Wong, P. W. K., O'Flynn, M. E., Inouye, T.: Micromethods for measuring phenylalanine and tyrosine in serum. Clin. Chem. *10,* 1098—1104 (1964).

Wurtman, R. J., Fernstrom, J. D.: L-Tryptophan, L-tyrosine, and the control of brain monoamine biosynthesis. In: Perspectives in Neuro-pharmacology (*Snyder, S. H.,* ed.), Vol. 1, pp. 143—193. Oxford: University Press. 1972.

Authors' address: Dr. *J.-M. Gaillard,* Clinique Psychiatrique de l'Université de Genève, Bel-Air, CH-1225 Chêne-Bourg, Switzerland.

Journal of Neural Transmission, Suppl. 15, 197—207 (1979)

Distribution of Tryptophan and Tyrosine in Unipolar Affective Disorders as Defined by Multicompartmental Analysis

D. M. Shaw, R. Blazek, S. F. Tidmarsh, G. J. Riley, A. L. Johnson, and **A. Michalakeas**

Biochemical Psychiatry Laboratory, Department of Psychological Medicine, Welsh National School of Medicine, Whitchurch Hospital, Cardiff, and M.R.C. Statistical Research and Services Unit, University College Hospital Medical School, London, U.K.

With 3 Figures

Summary

As requirements for tryptophan for synthesis of protein and 5-hydroxytryptamine were comparable in rat brain, during depletion of tryptophan there could be competition between the two pathways for the amino acid. This implied that tryptophan should be rate-limiting for protein synthesis and this was found in the short term when concentrations of the amino acid were reduced in rats.

Multicompartmental studies of tryptophan and tyrosine in controls and patients subject to unipolar depression defined two main pools of the amino acid provisionally assigned to extracellular and intracellular spaces.

For tyrosine, mean values for the extracellular space were comparable to those of controls. The concentration of tyrosine was low in the intracellular space in both depressed and recovered patients, but the raised fractional clearance rates for this compartment during depression had returned to normal on remission.

Plasma tryptophan concentrations were significantly reduced in depression with intermediate values after recovery. This suggested that the procedure used may have been mildly stressful and that this had evoked an idiosyncratic response to the stress in the depressed patients, which was characterized by inability to maintain concentrations of this amino acid in plasma.

The findings for both amino acids may have a bearing on the aetiology of unipolar affective disorder.

Noradrenaline and 5-hydroxytryptamine (5-HT) are two monoamines which act as neurotransmitters and which are thought to be implicated in the aetiology of affective disorders. This has led to studies of their synthetic pathways and their amino acid precursors, tyrosine and tryptophan respectively. The basis of this interest lays in the fact that if there were a central dysfunction involving the amines, this could be the result of a disturbance in the supply of amino acids to the brain.

In the search for a peripheral change in the distribution or utilization of these amino acids which might lead to such an abnormality, tryptophan seemed the more promising lead on the grounds that there was more evidence for a relationship between peripheral metabolism of tryptophan and 5-HT function than for tyrosine and catecholamines. It is the only essential amino acid bound to protein in plasma (*McMenamy* and *Oncley*, 1958) giving a concentration in non-protein bound form in the blood considerably lower than that of other amino acids and the supply of tryptophan to the brain depends in part on its concentration in plasma (*Pratt*, 1976; *Wurtman* and *Fernstrom*, 1974). Finally, the enzyme responsible for the first metabolic step in the conversion of tryptophan to 5-HT, tryptophan hydroxylase, is unsaturated under normal conditions (*Green* and *Sawyer*, 1966; *Moir* and *Eccleston*, 1968); consequently the concentration of the amino acid substrate in the brain is an important determinant of the rate of production of 5-HT.

These considerations tended to look only at the vulnerability of amine metabolism to an inadequate supply of the amino acid, but the situation may be more complex than this. The figures for turnover of tryptophan in 5-HT in the raphé region has been estimated by one group as 25.6 nmol/g/hour (*Neckers* and *Meek*, 1976). This figure is of comparable magnitude to that for turnover of tryptophan in protein synthesis in the brain of 25.9 nmol/g/hour (this communication). The fact, therefore, that the requirements for tryptophan of these two pathways were similar, raised the possibility that there could be competition for tryptophan between them in serotoninergic neurones.

For this to be shown to occur, other criteria have to be fulfilled. One is that tryptophan must be rate-limiting for protein synthesis in the brain, and also the rates of 5-HT and protein synthesis should be inversely related.

As an initial step, we have done experiments designed to find out if tryptophan is rate-limiting for synthesis of protein in the brain by reducing its concentrations in brain and plasma in rats and following the incorporation of the labelled amino acid (*Blazek* and *Shaw*, 1978).

The animals were given a constant intravenous infusion of radioactive tryptophan of high specific activity over a period of six hours. The ratio of:

$$\frac{\text{dpm incorporated/gm brain}}{\text{specific activity of tryptophan in brain homogenate}}$$

was plotted against time. This gave a straight line passing through the origin and a slope indicating a rate of incorporation of tryptophan into brain protein of 25.9 ± 2.22 nmoles/g/hour or 7.1 % per day.

The ratio measured at 3 hours infusion was used to compare protein synthesis rates in treated and control animals.

Tryptophan levels were reduced with α-methyl tryptophan in a dose of 50 mg/kg body weight intraparenterally (*Sourkes*, 1974). Low tryptophan concentrations were maintained for 6 hours, 24 hours or 1 week by a single injection and thereafter by weekly injections (Fig. 1). This reduced the rate of incorporation of tryptophan into plasma proteins throughout. The corresponding effect for brain proteins took place only over a limited period, after which protein synthesis rates "recovered", despite continuing low concentrations of tryptophan in the brain (Fig. 2).

Infusion of tryptophan had no effect on the incorporation of labelled amino acids into plasma or brain proteins of untreated animals, but in those which had received α-methyl tryptophan, tryptophan reversed the effect of the drug.

This was our first step in the study of the possible interactions of tryptophan deficiency with protein/amine synthetic pathways and we are going on to seek evidence for competition between them in areas rich in serotonergic neurones.

Thus the present data suggested that low levels of tryptophan had led to a reduction in the rates of synthesis of proteins followed by apparent "recovery". The most likely explanation for the "recovery" of protein synthesis in the brain was that it was due to metabolic readjustment. If this interpretation proved correct, the process might be a significant factor in humans if a transient suppression of protein synthesis in the brain were sufficient to disrupt function, or if the process were repetitive as in an intermittent reduction in tryptophan concentrations in plasma and brain, or if the recovery process were defective in some way.

In our studies of control subjects and patients with unipolar affective disorders we have concentrated on a search for changes in peripheral distribution and utilization of tryptophan or tyrosine which might secondarily affect the brain or which might be the indicants of other types of disturbance in amino acid metabolism. To

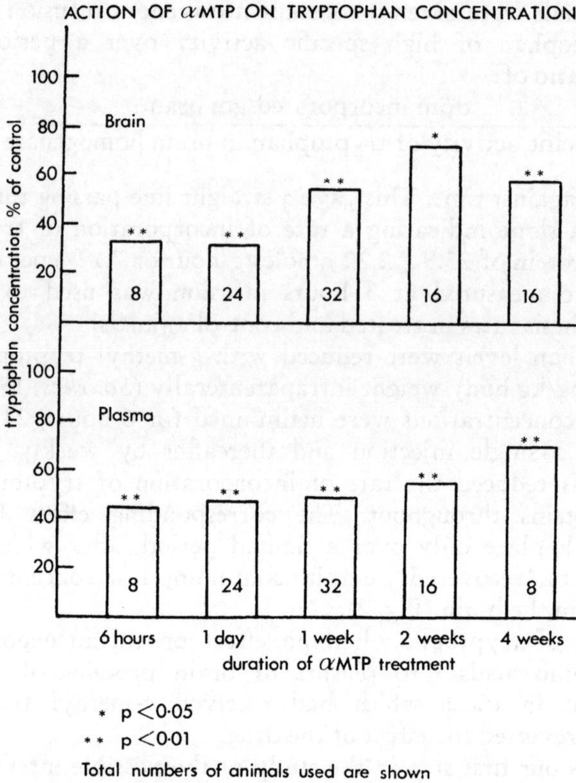

Fig. 1. Rats were injected with saline or 50 mg/kg α-MTP i.p. at weekly intervals. Tryptophan concentration was measured by dilution of a ^{14}C-tryptophan trace label after separation from α-MTP by paper chromatography. The results presented are derived from rats which had been anaesthetized and infused with ^{3}H-tryptophan for 3 hours. The numbers within each block represent the total number of rats used. Differences between controls and α-MTP treated rats were compared by the Student's t-test; * $p < .05$, ** $p < .01$

gain a more comprehensive view than that obtained from simpler studies of plasma levels, etc., we have used multicompartmental analysis.

Multicompartmental analysis is a useful method, with the advantage that it can give information on metabolic compartments in a way which is not possible with other currently available in vivo techniques.

Its limitations are at the mathematical/biological interface. Even with the very simple model used here, it is difficult to be certain that the theoretical model chosen is an accurate representation of the biological processes. Another possible limitation is that it gives an

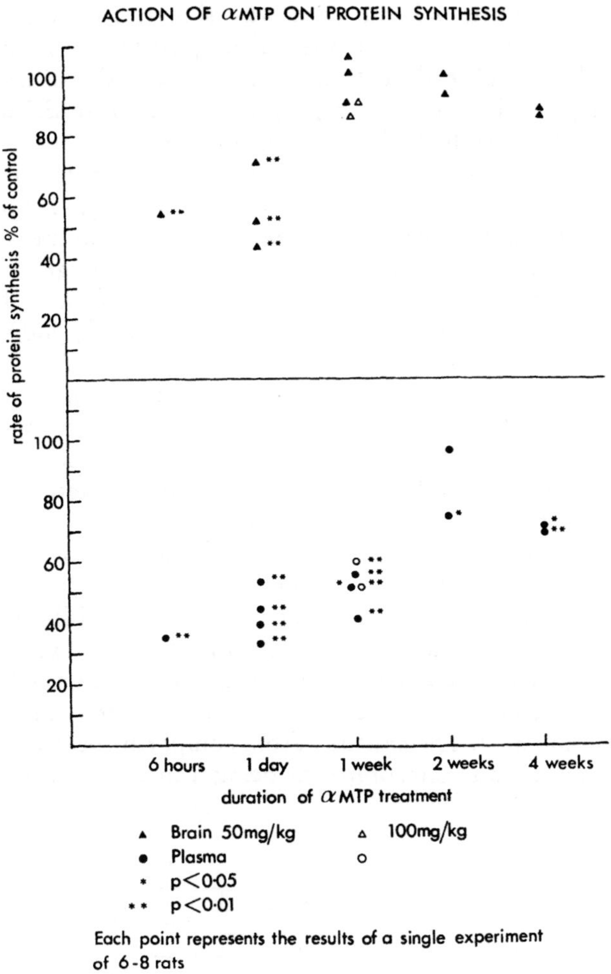

Fig. 2. Rats were injected i.p. with saline or α-MTP at a dose of 50 mg/kg (closed symbols) or 100 mg/kg (open symbols) at weekly intervals. They were anaesthetized and infused with ^{3}H-tryptophan. The ratio of the dpm incorporated per gram of tissue to the specific activity of the homogenate at three hours was taken as an index of the rate of protein synthesis. Each symbol (▲△ brain, ●○ plasma) represents the results of one experiment consisting of a total of 6—8 rats. Differences between controls and α-MTP treated rats were compared by the Student's t-test; * $p < .05$, ** $p < .01$

overall view which does not distinguish the individual contributions of cells, tissues and organs.

Thus data from multicompartmental analysis, whether giving "normal" values or suggesting differences between groups must be regarded as provisional, but in a way similar to the results of, for instance, studies of brain tissue taken post-mortem, may provide heuristically useful leads as to the existencc and site of abnormalities. The strategy of its use is to employ it as a probe to delineate change which must then be checked and verified by more direct techniques.

The method involved the injection of a tracer dose of a tritium-labelled tryptophan or tyrosine, the assay of change in specific activity of the amino acid in plasma with time, and measurements of the concentration of the amino acid in plasma throughout the 200 or 250 min of the experiment (*Shaw, Tidmarsh, Johnson, Michalakeas, Riley, Blazek*, and *Francis*, 1978; *Shaw, Tidmarsh, Johnson, Michalakeas, Riley, Blazek, Mac Sweeney, Francis*, and *Hewland*, 1978). Reliable data were not possible in the first five minutes after the injection of the labelled amino acid because rate of change of specific activity in plasma was rapid compared with the time needed to take a sample. From 10 to 200 or 250 min the specific activity curve was described by two exponentials in all subjects.

These data could be fitted adequately to a two compartmental model (Fig. 3). By definition, as the site of sampling, compartment 1 included plasma, and because of the rapidity of equilibration of small molecules between plasma and interstitial fluid, extended to the interstitial space. If compartment 1 was largely extracellular, it was argued that compartment 2 probably represented an intracellular pool of amino acid. The exit flux from compartment 2 was essentially one way and so was interpreted as being a metabolic transformation.

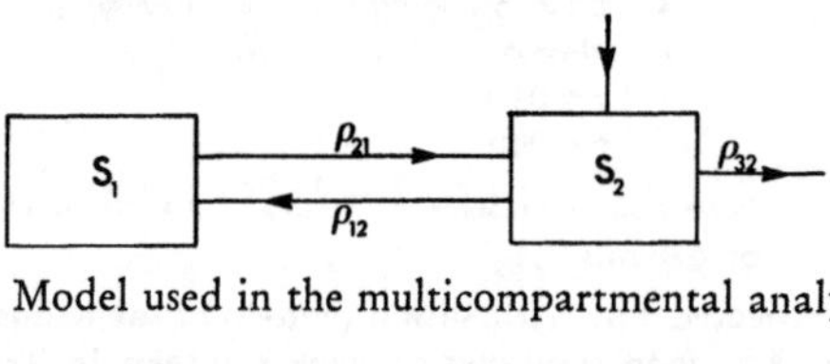

Fig. 3. Model used in the multicompartmental analysis

This provisional designation of the two compartments gave no information on the site of entry of "cold" amino acid presumably from protein degradation. The site of protein degradation had to be intracellular but under certain circumstances entry of amino acid to the system might appear as a flux into the extracellular space. It was thought that the conditions required for this to occur would be unlikely to be present, particularly in the fasting state, so the model shown with the designation of the two compartments to extra- and intracellular spaces was considered to be the most appropriate and was used in the analyses.

We assayed concentrations of amino acid in plasma and from this and the other data calculated the various fractional clearance rates ($min^{-1} \times 10^{-3}$), the concentration in compartment 2 (μM) and the corresponding fluxes representing rates of transfer of material between and through the compartments ($\mu M \times min^{-1}$).

All values for S_1 were concentrations in plasma which in theory might have led to difficulty in interpretation for tryptophan where there was binding to albumin in plasma and presumably in the interstitial space.

In practice, equilibration between protein bound and non-bound forms was so rapid (*Riley*, unpublished observation) that changes in binding were unlikely to have much, if any, effect on the rate of transport of the labelled amino acid.

This problem did not enter into our interpretations to any extent because, although changes in the binding of tryptophan have been reported in unipolar affective disorder (*Coppen* and *Wood*, 1978), we have been unable to find any alteration in the properties or concentration of binding protein or in the proportions of protein-bound and non-bound forms in a similar group of patients to the ones reported here (*Riley* and *Shaw*, 1976; *Riley* and *Shaw*, 1978).

Although S_2 was expressed as a concentration, in the absence of a measure of the volume of this compartment, it was taken as a comparative value, not as an absolute measure of the concentration of the amino acid in the second compartment. This was valid only if there was no change in the relative volumes of the two compartments. Although there was evidence for some loss of total body water in depression which was replaced after recovery, the relative sizes of the extra- and intracellular spaces remained remarkably constant (*Coppen* and *Shaw*, 1963; *Coppen*, *Shaw*, *Malleson*, and *Costain*, 1966).

In Table 1 are the details of the sample of controls and unipolar depressives included in this study, and in Tables 2 and 3 data for tyrosine and tryptophan. For tyrosine all the parameters associated with compartment 1 were similar to controls in both depressed and recovered patients and this included the concentrations of the amino acid in plasma, fractional clearance rates and the associated inter-compartmental fluxes. The abnormalities lay in compartment 2, as shown by reduced concentrations in patients during depression and after recovery. The depressed patients had an additional abnormality; raised fractional clearance rates (λ_{12} and λ_{32}) associated with compartment 2.

The findings for tryptophan were quite different. Although there appeared on first examination to be some possible differences in the

Table 1. *Description of sample of unipolar patients and controls*

			Males	Females	Total	Age (mean ± S.D.) Males	Age (mean ± S.D.) Females
Tyrosine study	Controls		10	12	22	45.7 ± 10.4	45.4 ± 9.1
	Unipolar affective disorder	recovered	4	10	14	55.0 ± 7.3	59.3 ± 8.8
		depressed	5	14	19	58.6 ± 11.9	48.7 ± 14.9
Tryptophan study	Controls		10	10	20	44.4 ± 6.5	42.5 ± 5.9
	Unipolar affective disorder	recovered	9	11	20	54.2 ± 6.2	55.2 ± 11.1
		depressed	7	15	22	60.0 ± 5.7	52.9 ± 14.4

Table 2. *Tyrosine: Parameters associated with Compartment 2 (means ± S.E.)*

	N	Concentration Compartment 2 (μM)	Fractional clearance rate λ_{12} ($min^{-1} \times 10^{-3}$)	Fractional clearance rate λ_{32} ($min^{-1} \times 10^{-3}$)
Control	22	285.2 ± 16.6	10.4 ± 0.7	12.1 ± 0.7
Unipolar affective disorder (depressed)	19	225.6 ± 17.8**	13.2 ± 0.8***	14.5 ± 0.7*
Unipolar affective disorder (recovered)	14	217.8 ± 20.8**	11.4 ± 0.9	12.8 ± 0.9

Significance of difference from controls: * $p < 0.05$, ** $p < 0.02$, *** $p < 0.01$.

Table 3. *Concentrations of tryptophan in plasma (mean ± S.E.)*

	N	Concentration of tryptophan in plasma (μM)	Significance of difference from controls
Controls	20	63.0 ± 2.2	—
Unipolar affective disorder	42	54.1 ± 1.5	$p < 0.005$
Unipolar affective disorder (depressed)	22	51.4 ± 2.1	$p < 0.001$
Unipolar affective disorder (recovered)	20	56.9 ± 2.2*	n.s.

* Not significantly different from depressed group.

parameters associated with compartment 2, none of these was significantly different from control values, at least in the group as a whole. (There was some evidence in males only that there was an extension of abnormalities in tryptophan to parameters other than the one outlined below, but these data have not been included here.)

The main change for tryptophan, at least in the group as a whole, lay in the extracellular space as represented by the mean plasma value which was significantly reduced in depressed patients. That of recovered depressives lay between the mean plasma concentrations of depressed and control individuals being significantly different from neither.

We had not seen a comparable fall in tryptophan concentrations in plasma in a similar group of unipolar patients called upon at short notice to give just a single sample of blood.

The difference may have been in the experimental conditions. The multicompartmental analysis was a complex technique which may have been mildly stressful to the patient in anticipation and perhaps in experience, despite preceding explanations. If so, then the unipolar depressives had demonstrated an idiosyncratic response to stress which had been characterized by a failure to maintain concentrations of tryptophan in plasma. Since the change was in plasma, this could have a bearing on the supply of the amino acid to the brain and conceivably, as it was evoked by stress, the process might occur only intermittently.

In conclusion, therefore, we have shown that tryptophan can be rate-limiting for protein synthesis in the brain in rats, at least in the short term, a finding which may be relevant for human disorders.

Multicompartmental studies have shown systematic differences in tyrosine between controls, patients with unipolar depression and patients who have recovered from this illness, which included illness-dependent and illness-independent changes. The latter were tentatively interpreted in terms of alterations of tyrosine content in the intracellular space.

Similar data for tryptophan suggested no changes for the intracellular compartment in the group as a whole but there was an illness-dependent, probably stress-induced, decrease in plasma concentrations of this amino acid. A possible effect of this could be in the supply of tryptophan to the brain.

Acknowledgements

We wish to thank the medical and nursing staffs of Whitchurch and West Park Hospitals for their help throughout the study and in particular Sister I. Hockley. We thank Mrs. D. Barrett for her secretarial assistance.

The Biochemical Psychiatry Laboratory is supported on an MRC Programme Grant.

References

Blazek, R., Shaw, D. M.: Tryptophan availability and brain protein synthesis. Neuropharmacology (in press).

Coppen, A., Shaw, D. M.: Mineral metabolism in melancholia. Lancet *ii,* 1439—1444 (1963).

Coppen, A., Shaw, D. M., Malleson, A., Costain, R.: Mineral metabolism in mania. Brit. Med. J. *1,* 71—75 (1966).

Coppen, A., Wood, K.: Tryptophan and depressive illness. Psychol. Med. *8,* 49—57 (1976).

Green, H., Sawyer, J. L.: Demonstration, characterization and assay procedure of tryptophan hydroxylase in the rat brain. Anal. Biochem. *15,* 53—64 (1966).

McMenamy, R. H., Oncley, J. L.: The specific binding of L-tryptophan to serum albumin. J. Biol. Chem. *233,* 1436—1447 (1958).

Moir, A. T. B., Eccleston, D. J.: The effect of precursor loading in the cerebral metabolism of 5-hydroxyindoles. J. Neurochem. *15,* 1093 to 1108 (1968).

Neckers, L. M., Meek, J. L.: Measurement of 5-HT turnover rate in discrete nuclei of rat brain. Life Sciences *19,* 1579—1584 (1976).

Pratt, O. E.: Transport Phenomena in the Nervous System (*Levi, G., Battistin, L., Lajtha, A.,* eds.), pp. 55—75. New York: Plenum Press. 1976.

Riley, G. J., Shaw, D. M.: Total and non-bound tryptophan in unipolar illness. Lancet *ii,* 1249 (1976).

Riley, G. J., Shaw, D. M.: In preparation (1978).

Shaw, D. M., Tidmarsh, S. F., Johnson, A. L., Michalakeas, A. C., Riley, G. J., Blazek, R., Francis, A. F.: Multicompartmental analysis of amino acids. II. Tryptophan in affective disorders. Psychol. Med. (in press, 1978).

Shaw, D. M., Tidmarsh, S. F., Johnson, A. L., Michalakeas, A. C., Riley, G. J., Blazek, R., Mac Sweeney, D. A., Francis, A. F., Hewland, R.: Multicompartmental analysis of amino acids. III. Tyrosine in affective disorders. Psychol. Med. (in press, 1978).

Sourkes, T. L.: Aromatic amino acids in the brain. In: Ciba Foundation Symposium 22 (new series), pp. 361—380. Amsterdam: Elsevier. 1974.

Wurtman, R. J., Fernstrom, J. D.: Control of brain serotonin by the diet. Adv. in Neurol. *5,* 19—29 (1974).

Authors' address: Dr. *D. M. Shaw,* Whitchurch Hospital, Cardiff, CF4 7XB, U.K.

[illegible] tryptophan and 5-HT [illegible] singular reference [illegible]

The [illegible] thanks [illegible] [illegible] supported [illegible] grant.

References

[illegible]: Tryptophan availability and [illegible] synthesis [illegible] (in press).

[illegible]

[illegible], Med. [illegible] (1966).

[illegible]

[illegible], characterization, and [illegible] of tryptophan hydroxylase in the rat brain [illegible] (1969).

[illegible]: The [illegible] availability of [illegible] tryptophan to [illegible] albumin [illegible] (1969).

[illegible]

[illegible] (1969).

[illegible]: [illegible] New York: [illegible] Press.

[illegible]: Free and bound tryptophan [illegible]

[illegible]

[illegible]

[illegible]: [illegible] tryptophan [illegible]

[illegible]

[illegible]

[illegible]

[illegible]: Control of brain serotonin by the [illegible] (1971).

Author's address: Dr. [illegible], [illegible]

Journal of Neural Transmission, Suppl. 15, 209—216 (1979)

Serotonergic Mechanisms in Myoclonus

J. H. Growdon

Department of Neurology, Tufts-New England Medical Center Hospital, Boston, Massachusetts, U.S.A.

Summary

Post-hypoxic intention myoclonus is a specific myoclonic syndrome in which central serotonergic tone may be deficient. Tryptophan and 5-hydroxytryptophan administration to patients with post-hypoxic intention myoclonus increases pre-existing low levels of 5-hydroxyindoleacetic acid levels in the cerebrospinal fluid, and also suppresses myoclonus. Serotonin precursor administration does not help all patients with myoclonus and may actually worsen some myoclonic syndromes, including those secondary to lipid storage diseases. Treatments that alter serotonin metabolism can also produce myoclonus in experimental animals but their relevance to myoclonic syndromes in humans remains uncertain.

Myoclonus is a descriptive term for quick involuntary non-repetitive muscular contractions. Myoclonus occurs in a wide variety of neurologic diseases which, collectively, are called the myoclonias (*Bonduelle*, 1968). Myoclonic syndromes may differ according to etiology (*Swanson et al.*, 1962), clinical manifestations (*Halliday*, 1967 a), or electrophysiologic characteristics (*Halliday*, 1967 b; *Hallett et al.*, 1977). Recent pharmacologic studies indicate that there are biochemical differences among the myoclonias as well. Thus, serotonin precursors improved patients with post-hypoxic intention myoclonus (PHIM), whereas they may worsen patients with other myoclonic syndromes.

Post-hypoxic intention myoclonus is a specific myoclonic syndrome that results from transient brain anoxia, such as occurs during a cardiorespiratory arrest (*Lance* and *Adams*, 1963). Action myoclonus is the most incapacitating feature of the illness, although residual defects may also include varying amounts of muscle weakness and incoordination. Myoclonus occurs during volitional movements and the muscles are usually quiet "at rest", although sensory stimuli such as flashes of lights, loud noises, or an unexpected touch may also produce myoclonus in some patients. Cognitive abilities are usually preserved, but speech is often slurred and halting due to pharyngeal or diaphragmatic myoclonus.

Evidence for Serotonin Deficiency in PHIM

A single report on the post-mortem examination of a brain from a patient with PHIM revealed widespread anoxic changes (*Castaigne et al.*, 1964); no biochemical measurements were made. Evidence that a serotonin deficiency causes PHIM is based on clinical studies in which investigators found that (1) levels of 5-HIAA in the CSF of patients with PHIM were significantly lower than in control subjects, and (2) tryptophan and 5-HTP administration elevated 5-HIAA levels in the CSF and also suppressed myoclonus (Table 1).

Table 1. *Effect of 5-HTP administration on myoclonus in patients with PHIM. A summary of 9 reports*

Investigator	Number of patients	Results
Lhermitte et al. (1971)	1	Decreased Myoclonus
Lhermitte et al. (1972)	1	Decreased Myoclonus
Guilleminault et al. (1973)	1	Decreased Myoclonus
Van Woert and *Sethy* (1975)	3	Decreased Myoclonus
Chadwick et al. (1975)	3	Decreased Myoclonus
DeLean et al. (1976)	2	Decreased Myoclonus
Growdon et al. (1976)	2	Decreased Myoclonus
Van Woert et al. (1976)	12	Decreased Myoclonus in 9
Chadwick et al. (1977)	5	Decreased Myoclonus in 3

5-HTP Administration to Patients With Myoclonic Syndromes

In order to determine whether there were serotonin deficiencies in other myoclonic syndromes, *Growdon et al.* (1976) gave oral doses of 5-HTP in conjunction with carbidopa to 2 patients with PHIM, 2 patients with idiopathic action myoclonus, 1 patient with a congenital encephalopathy and action myoclonus, 1 patient with myoclonic epilepsy, and 2 patients with lipid storage disease

Table 2. *Effects of 5-HTP administration to 8 patients with different myoclonic syndromes*

Patient	Etiology of myoclonus	Maximum dose of 5-HTP (mg/day)	Clinical effects	Side effects
1	Post-hypoxic	1050	Decreased Myoclonus	Nausea, Vomiting Diarrhea
2	Post-hypoxic	1200	Decreased Myoclonus	Nausea, Vomiting Increased Seizures Toxic Psychosis
3	Idiopathic	500	Decreased Myoclonus	Nausea, Anorexia
4	Congenital Encephalopathy	900	No Change	None
5	Myoclonic Epilepsy	400	Increased Myoclonus	Nausea, Vomiting Depression
6	Idiopathic	1000	Increased Myoclonus	None
7	Lipid Storage	400	Increased Myoclonus	Nausea, Vomiting Increased Seizures
8	Lipid Storage	75	?	Increased Seizures

(Adapted from *Growdon et al.*, 1976.)

(Table 2). The initial dose of 5-HTP was 1 mg/kg/day; the maximum dose ranged between 5—15 mg/kg/day in divided doses. The dose of carbidopa, given to inhibit aromatic amino acid decarboxylase in the periphery, remained constant throughout the study. A patient's response to treatment was monitored by periodic clinical examinations, combined surface EMG and accelerometer recordings, and analyses of movies.

The 2 patients with PHIM improved greatly while taking 700 to 1000 mg of 5-HTP per day. The case history of patient 2 and her response are summarized below.

Mrs. C. had an acute myocardial infarction with pulmonary edema; she had a respiratory arrest with loss of blood pressure and pulse shortly after arrival in the hospital emergency room. She was promptly intubated and blood pressure, pulse, and ventilation were restored within an estimated 3—5 min. Following resuscitation, she was comatose with decorticate posturing and had frequent intermittant generalized seizures. On the second hospital day she was still comatose and developed spontaneous myoclonus in the arms, legs, and trunk. She regained consciousness by the third hospital day; spontaneous myoclonus subsided over the course of the next few days but action myoclonus became more prominent. On examination 3 months after the cardiorespiratory arrest her mind was clear and there was no aphasia. She had severe facial myoclonus whenever she pursed her lips, protruded her tongue, or closed her eyes. Myoclonus also prevented clear speech and she was often unable to whisper her name. She lay quietly in bed, but unexpected tactile stimuli elicited myoclonus in the body part that was touched. Action myoclonus was so severe that she could not comb her hair, feed herself, or even hold a pencil to write, despite good strength in her arms. The deep tendon relexes were hyperactive and there were bilateral extensor plantar responses. She could not stand or walk, and attempts to elevate her legs against gravity produced severe myoclonus that spread throughout the abdominal musculature. When 5-HTP was added to her drug regimen, the myoclonus was greatly suppressed. She was able to say complete sentences smoothly, and the facial myoclonus decreased. She regained the ability to dress herself, feed herself, and even to write. Although the myoclonus was not completely suppressed, she was able to walk between the parallel bars for the first time since her illness.

Although patient 3 also improved during 5-HTP administration as well, 5-HTP failed to suppress myoclonus in the other patients, and the myoclonus actually increased in 3 (Table 2).

Side-effects from 5-HTP limited its utility, even in those patients who improved during its ingestion. Patients 1, 2, 3, 5, and 7 all had significant nausea, vomiting, and diarrhea despite the slow gradual administration of 5-HTP and its relatively low effective dose made possible by the concurrent use of carbidopa. In addition, patient 2

(Mrs. C.) developed a toxic psychosis while taking 1200 mg/day of 5-HTP and patient 5 become depressed while taking 400 mg/day. Both patients with lipid storage disease had an increased number of seizures while taking 5-HTP; these subsided when 5-HTP was discontinued.

Discussion

The results of this study, and those from other studies (*Lhermitte et al.*, 1971, 1972; *Guilleminault et al.*, 1973; *Van Woert et al.*, 1975; *Chadwick et al.*, 1975; *DeLean et al.*, 1976; *Growdon et al.*, 1976; *Van Woert et al.*, 1977; *Chadwick et al.*, 1977) indicate that 5-HTP administration can suppress myoclonus in patients with PHIM, presumably by increasing brain levels of serotonin. Clonazepam, a benzodiapine, may be equally effective in suppressing action myoclonus (*Chadwick et al.*, 1977); clonazepam is also reported to increase whole brain levels of serotonin (*Jenner et al.*, 1975). Tryptophan administration, alone or in combination with a monoamine oxidase inhibitor (MAO-I), also suppressed myoclonus in some patients with PHIM (*DeLean et al.*, 1976; *Chadwick et al.*, 1977), but not in others (*Van Woert et al.*, 1975). These data led to the hypothesis that PHIM results from a central serotonergic deficiency, in the same sense that Parkinson's disease is a result of dopamine deficiency. Tryptophan and 5-HTP both cross the blood-brain barrier, accelerate serotonin synthesis, and restore serotonergic tone. Although the clinical results are consistent with the view that increased serotonergic transmission can suppress PHIM, other neurotransmitters may be involved as well. Serotonin can be synthesized from 5-HTP in any cell that contains aromatic amino acid decarboxylase, including neurons that normally synthesize catecholamines. 5-HTP administration could therefore displace catecholamines from their usual sites and thereby produce effects not normally associated with serotonin release. The low level of HVA in the CSF of patients treated with 5-HTP is consistent with this possibility. Tryptophan administration might be a more specific treatment since tryptophan hydroxylase, the enzyme that catalyzes the conversion of tryptophan to 5-HTP and limits the rate of serotonin synthesis, is exclusively located within serotonergic neurons. Tryptophan administration does increase brain levels of serotonin (*Fernstrom* and *Wurtman*, 1971), but it also increases the ratio of the plasma concentration of tryptophan to the concentration of the sum of the neutral amino acids that compete with it for uptake into the brain (*Fernstrom* and *Wurtman*, 1972). Thus by suppressing the transport of the other neutral amino acids, including the catecholamine precursor tyrosine, large doses of tryptophan can slow catecholamine synthesis.

Although most patients with PHIM improved during tryptophan or 5-HTP administration, patients with some other myoclonic syndromes did not (*Guilleminault et al.*, 1973; *Chadwick et al.*, 1977); myoclonus actually increased in some patients, including those with lipid storage disease (*Growdon et al.*, 1976). These variable clinical responses to 5-HTP indicate the heterogeneity of serotonergic mechanisms among the myoclonias. In order to examine these serotonergic responses further, investigators have studied myoclonus in experimental animals. *Grahame-Smith* (1971) reported that tryptophan in conjunction with an MAO-I induced clonic fits in rats, and *Klawans et al.* (1975) reported that 5-HTP alone induced myoclonic shocks in immature guinea pigs. *Stewart et al.* (1976) and *Trulson et al.* (1976) independently and concurrently found that they could produce myoclonus by administering small doses of 5-HTP to rats who had previously received the serotonin neurotoxin 5, 7-dihydroxytryptamine; this syndrome could be blocked by the administration of serotonin antagonists. Drugs that act as serotonin antagonists also blocked the postural myoclonic syndrome that followed p-chloroamphetamine injections (*Growdon*, 1977, 1978) or fenfluoramine administration (*Trulson* and *Jacobs*, 1976). Thus, most treatments that increase levels of serotonin in the brain, or that potentiate its post-synaptic effects, produce myoclonus in experimental animals. In contrast, *Van Woert* (1978) found that 5-HTP could block the spontaneous and stimulus-sensitive myoclonus produced in rats by the oral administration of 600 mg/kg of p,p′-DDT. These experimental data indicate that altered serotonin metabolism can cause myoclonus in animals as well as in humans. There are no animal models of PHIM however, and the extent to which myoclonic syndromes in animals correlate with specific myoclonias in humans remains unknown.

References

Bonduelle, M.: The Myoclonias. In: Handbook of Clinical Neurology (*Vinken, P. J., Bruyn, G. W.*, eds.), Vol. 6, Ch. 29, pp. 761—781. Amsterdam: North Holland. 1968.

Castaigne, P., Cambier, J., Escourolle, R., Cathala, H. P., Lacasble, R.: Observation anatomo-clinique d'un syndrome myoclonique post-anoxique. Rev. Neurol. *111*, 60—73 (1964).

Chadwick, D., Jenner, P., Harris, R., Reynolds, E. H., Marsden, C. D.: Manipulation of brain serotonin in the treatment of myoclonus. Lancet *2*, 434—435 (1975).

Chadwick, D., Hallett, M., Harris, R., Jenner, P., Reynolds, E. H., Marsden, C. D.: Clinical, biochemical, and physiological features distinguishing myoclonus responsive to 5-hydroxytryptophan, tryptophan with a monoamine oxidase inhibitor, and clonazepam. Brain *100,* 455—487 (1977).

DeLean, J., Richardson, J. C., Hornykiewicz, O.: Beneficial effects of serotonin precursors in postanoxic action myoclonus. Neurology *26,* 863—868 (1976).

Fernstrom, J. D., Wurtman, R. J.: Brain serotonin content: physiological regulation by plasma neutral amino acids. Science *178,* 414—416 (1972).

Guilleminault, C., Tharp, B. R., Cousin, D.: HVA and 5-HIAA measurements and 5-HTP trials in some patients with involuntary movements. J. Neurol. Sci. *18,* 435—441 (1973).

Grahame-Smith, D. G.: Studies *in vivo* on the relationship between brain tryptophan, brain 5-HT synthesis, and hyperactivity in rats treated with monoamine oxidase inhibitor and L-tryptophan. J. Neurochem. *18,* 1053—1066 (1971).

Growdon, J. H., Young, R. R., Shahani, B. T.: L-5-hydroxytryptophan in treatment of several different syndromes in which myoclonus is prominent. Neurology *26,* 1135—1140 (1976).

Growdon, J. H.: Postural changes, tremor, and myoclonus in the rat immediately following injections of p-chloroamphetamine. Neurology *27,* 1074—1077 (1977).

Growdon, J. H.: Changes in motor behavior following the administration of serotonin neurotoxins. Annals N.Y. Acad. Sci. *305,* 510—523 (1978).

Hallett, M., Marsden, C. D.: Ballistic movement overflow myoclonus. A form of essential myoclonus. Brain *100,* 299—312 (1977).

Halliday, A. M.: The clinical incidence of myoclonus. In: Modern Trends in Neurology (*Williams, D.,* ed.), Vol. 4, pp. 69—105. London: Butterworth. 1967.

Halliday, A. M.: The electrophysiological study of myoclonus in man. Brain *90,* 241—284 (1967).

Jenner, P., Chadwick, D., Reynolds, E. H., Marsden, C. D.: Altered 5-HT metabolism with clonazepam and diphenylhydantoin. J. Pharm. Pharmacol. *27,* 707—710 (1975).

Klawans, H. L., Goetz, C., Weiner, W. J.: 5-Hydroxytryptophan-induced myoclonus in guinea pigs and the possible role of serotonin in infantile myoclonus. Neurology *23,* 1234—1240 (1973).

Lance, J. W., Adams, R. D.: The syndrome of intention or action myoclonus as a sequel to hypoxic encephalopathy. Brain *86,* 111—136 (1963).

Lhermitte, F., Peterfalvi, M., Marteau, R., Gazengel, J., Serdaru, M.: Pharmacological analysis of a case of postanoxic intention and action myoclonus. Rev. Neurol. *124,* 21—31 (1971).

Lhermitte, F., Marteau, R., Degos, C. F.: Analyse pharmacologique d'un nouveau cas de myoclonies d'intention et d'action post-anoxiques. Rev. Neurol. *126,* 107—114 (1972).

Stewart, R. M., Growdon, J. H., Cancian, D., Baldessarini, R. J.: 5-hydroxytryptophan-induced myoclonus: increased sensitivity to serotonin after intracranial 5, 7-dihydroxytryptamine in the adult rat. Neuropharm. *15*, 449—455 (1976).

Swanson, P. D., Luttrell, C. N., Magladery, J. W.: Myoclonus—a report of 67 cases and a review of the literature. Medicine *41*, 339—356 (1962).

Trulson, M. D., Jacobs, B. L.: Behavioral evidence of the rapid release of CNS serotonin by PCA and fenfluoramine. Eur. J. Pharmacol. *36*, 149—154 (1976).

Trulson, M. E., Eubanks, E. E., Jacobs, B. L.: Behavioral evidence for supersensitivity following destruction of central serotonergic nerve terminals by 5, 7-dihydroxytryptamine. J. Pharmacol. Exp. Ther. *198*, 23—32 (1976).

Van Woert, M. H., Hwang, E. C.: Myoclonus: biochemical and pharmacological approaches. In: Clin. Neuropharm. Vol. 3 (*Klawans, H. L.*, ed.). New York: Raven Press. 1978.

Van Woert, M. H., Sethy, V. H.: Therapy of intention myoclonus with L-5-hydroxytryptophan and a peripheral decarboxylase inhibitor, MK 486. Neurol. *25*, 135—140 (1975).

Van Woert, M. H., Jutkowitz, R., Rosenbaum, D., Bowers, M. B., jr.: Serotonin and myoclonus. Monogr. neural Sci. *3*, 71—80 (1976).

Van Woert, M. H., Rosenbaum, D., Howieson, J., Bowers, M. B., jr.: Long-term therapy of myoclonus and other neurologic disorders with L-5-hydroxytryptophan and carbidopa. N.E.J.M. *296*, 70—75 (1977).

Author's address: Dr. *J. H. Growdon*, Department of Neurology, Tufts-New England Medical Center Hospital, 171 Harrison Avenue, Boston, MA 02111, U.S.A.

Journal of Neural Transmission, Suppl. 15, 217—219 (1979)

Comment on the Methodology for Separating and Determining Free and Albumin-Bound Tryptophan Levels in Blood Samples

J. D. Fernstrom

Laboratory of Brain and Metabolism, Program in Neural and Endocrine Regulation, Massachusetts Institute of Technology, Cambridge, Massachusetts, U.S.A.

Tryptophan in blood circulates in two forms: free, and bound to serum albumin. About 75—80 % is normally bound to albumin; the exact amount depends on such factors as the species, when the animal or subject has last eaten (and the composition of the diet), and whether or not any drugs have been administered that might influence tryptophan binding to albumin (*e.g.*, salicylate, heparin). At present, the relationship between albumin binding and tryptophan availability to tissues (*e.g.*, brain) for uptake is a subject of much study. However, the interpretation of data from different laboratories is often complicated by the fact that a variety of methods is used to separate free from bound tryptophan in the blood, prior to its determination. Depending on the method used, the apparent level of free tryptophan can vary greatly. The purpose of this brief note is to offer suggestions for standardizing the method used to separate free tryptophan in the blood.

(1) Because the binding of tryptophan to albumin in blood is very sensitive to changes in pH and to some extent in temperature [1, 2], care must be taken to avoid post-sampling alterations in binding that follow when pH and temperature are allowed to fluctuate. Sample pH can readily be maintained by keeping the atmospheric CO_2 content at 5 % [1], such as by gassing sample tubes with 95 % N_2—5 % CO_2, beginning immediately after blood samples are collected, and continuing through to the final separation [3]. For this reason, techniques such as equilibrium dialysis, as originally described by *McMenamy* and *Oncley* [3] (involving the use of standard dialysis sack methodology), or the use of special mechanical

ultrafiltration devices equipped to regulate pCO_2 (*e.g.*, see *Bloxam et al.*, 1977) are advisable [3, 4, 5]. The method of ultrafiltration that employs the Amicon CF-50 Diaflo filtration cones is inferior, in that neither pH nor temperature can be maintained during the filtration process [3, 4, 5].

Equilibrium dialysis has the additional advantage that not only pH, but also temperature, can be held at the physiologic level (37 °C) throughout the dialysis process (3.5 hours dialysis at 37 °C) [1, 2, 3]. For samples containing high levels of non-esterified fatty acids (NEFA) and fats, incubation at 37 °C has been shown *not* to lead to *in vitro* rises in serum NEFA, and thus free tryptophan, due to lipemia [3]. This method is thus completely suitable for obtaining accurate estimates of the *in vivo* levels of free and bound tryptophan in blood. Provided that the special ultrafiltration devices [*e.g.*, 4, 5] are used at 37 °C, these too will probably yield filtrates containing physiologic levels of free tryptophan.

(2) The centrifugation of dialysis tubing containing serum or plasma samples [*e.g.*, 6] seems also to be of questionable value, unless it has been determined beforehand that the pore size has not been altered by distention of the dialysis tubing during centrifugation.

(3) Heparinized plasma is not suitable for the estimation of free and bound tryptophan levels in blood, inasmuch as heparin interferes with the binding of tryptophan to albumin [5, 7]. If plasma is to be used, each anticoagulant should first be tested to ascertain whether or not it too blocks tryptophan binding to albumin. EDTA has been reported by *McMenamy et al.* [2] *not* to interfere with tryptophan binding, and thus may be a suitable anticoagulant [5]. The use of serum avoids this problem altogether.

In summary, the preparation of an ultrafiltrate of serum or plasma for the estimation of free tryptophan levels should be performed under conditions of physiologic pH and temperature (*i.e.*, pH = 7.4; T° = 37 °C). Equilibrium dialysis is probably the *best* method for attaining these conditions [1, 2, 3]. Special filtration equipment that maintains pH [4, 5] and temperature may also give reliable values. However, centrifugation using either Amicon Diaflo filters, or dialysis tubing, is probably the *least* reliable method for preparing an ultrafiltrate of serum or plasma [8].

Once the ultrafiltrate has been prepared and is being assayed for tryptophan [9, 10], the inhibitory effects of light on the formaldehyde condensation reaction should carefully be avoided [11].

References

[1] *McMenamy, R. H., Oncley, J. L.:* The specific binding of L-tryptophan to serum albumin. J. Biol. Chem. *233,* 1436—1447 (1958).

[2] *McMenamy, R. H., Lund, C. C., Van Marcke, J., Oncley, J. L.:* The binding of L-tryptophan in human plasma at 37 °C. Arch. Biochem. Biophys. *93,* 135—139 (1961).

[3] *Madras, B. K., Cohen, E. L., Messing, R., Munro, H. N., Wurtman, R. J.:* Relevance of free tryptophan in serum to tissue tryptophan concentrations. Metabolism *23,* 1107—1116 (1974).

[4] *Bloxam, D. L., Hutson, P. H., Curzon, G.:* A simple apparatus for ultrafiltration of small volumes: application to the measurement of free and albumin-bound tryptophan in plasma. Analyt. Biochem. *83,* 130—142 (1977).

[5] *Baumann, P., Perey, M.:* The analysis of free tryptophan in human blood with the ultrafiltrator: a comparison with other methods. Clin. Chim. Acta *76,* 223—231 (1977).

[6] *Tagliamonte, A., Biggio, B., Vargiu, L., Gessa, G. L.:* Free tryptophan in serum controls brain tryptophan level and serotonin synthesis. Life Sci. *12,* 277—287 (1973).

[7] *Bourgoin, S., Faivre-Bauman, A., Hery, F., Ternaux, J. P., Hamon, M.:* Characteristics of tryptophan binding in the serum of the newborn rat. Biol. Neonate *31,* 141—154 (1977).

[8] *D'Elia, G., Lehmann, J., Raotma, H.:* Evaluation of the combination of tryptophan and ECT in the treatment of depression. II. Biochemical Analysis. Acta Psychiat. Scand. *56,* 319—334 (1977).

[9] *Denckla, W. D., Dewey, H. K.:* The determination of tryptophan in plasma, liver, and urine. J. Lab. Clin. Med. *69,* 160—169 (1967).

[10] *Bloxam, D. L., Warren, W. H.:* Error in the determination of tryptophan by the method of Denckla and Dewey. A revised procedure. Anal. Biochem. *60,* 621—625 (1974).

[11] *Lehmann, J.:* Light—a source of error in the fluorimetric determination of tryptophan. Scand. J. Clin. Lab. Invest. *28,* 49—55 (1971).

Author's address: Dr. *J. D. Fernstrom,* Laboratory of Brain and Metabolism, Program in Neural and Endocrine Regulation, Massachusetts Institute of Technology, Cambridge, MA 02139, U.S.A.

Journal of Neural Transmission, Suppl. 15, 221—226 (1979)

Methodological Problems in the Determination of Total and Free Plasma Tryptophan

G. Curzon

Department of Neurochemistry, Institute of Neurology, London, U.K.

There has been much discussion of the utility of plasma concentrations of tryptophan (total and free) and of amino acids competing with tryptophan for transport to the brain as predictors of brain tryptophan concentration. This issue now may be attaining resolution (*Curzon*, this volume; *Green*, 1978; *Wurtman*, this volume). It appears that large changes of competing amino acid concentrations (*e.g.* following food intake) have important effects on brain tryptophan while in other circumstances changes of plasma free tryptophan concentration (*e.g.* when drugs or physiological changes alter the binding of plasma tryptophan to albumin) also have considerable effects. In the past, different groups of workers have focused their attention on experimental situations in which one or other of the above influences were predominant and this has given an impression of a greater polarization of attitudes than was perhaps justified. Also much confusion has resulted from methodological problems and artefactual changes. The following note on methods for the determination of plasma tryptophan (total and free) and non-esterified fatty acid (NEFA) may therefore be helpful.

Tryptophan

The majority of laboratories determining plasma tryptophan uses variants of the method of *Denckla* and *Dewey* (1967). This depends on the conversion of tryptophan to a fluorphore, norharman, by heating under acid conditions with formaldehyde and ferric chloride. In the original method the ferric chloride is added to the 10 % trichloroacetic acid used for initial plasma deproteinization. Under these conditions, tryptophan is unstable especially in the presence of sunlight (*Lehmann*, 1971). As the rate of destruction is usually greater in standards than in plasma extracts and is independent of tryptophan concentration losses tend to lead to falsely high plasma tryptophan results and are not easily compensated for by the use of internal standards (*Bloxam* and *Warren*, 1974). These authors improved the method by omitting ferric chloride from the deproteinizing reagent and adding it immediately before the heating step. Breakdown of standards is then minimal as little destruction occurs during heating. However, even under these conditions it is as well not to expose standards and plasma supernatants to the acid for longer than necessary before heating as it is possible that ferric ions or other substances in deproteinized plasma or present as contaminants of the trichloroacetic acid may cause losses.

Using the Bloxam and Warren modification, recoveries of tryptophan added to normal rat or human plasmas are usually excellent. However, low recoveries may occur with pathological material *e.g.* plasmas from patients with chronic renal disease (*Sullivan et al.*, 1978). This did not occur if plasmas were first diluted 10—20-fold with water.

Free Tryptophan

The determination of free tryptophan in plasma requires its separation from the protein-bound fraction. A number of pitfalls are associated with this procedure and consequently many methods have been developed. These are discussed and compared by *Baumann* and *Perey* (1977) and by *Wood et al.* (1977). Most methods involve either dialysis or ultrafiltration. Dialysis methods (*Madras et al.*, 1974; *Bender et al.*, 1975) generally require simple equipment but are often cumbersome and slow. They also may be subject to error because of the disturbance of binding equilibria which necessarily occurs.

Ultrafiltration methods involve either centrifugation or gas pressure to force the ultrafiltrate through a membrane. Centrifugation through Amicon ultrafiltration cones (*Knott* and *Curzon*, 1972) is simple but has disadvantages. One problem is the deterioration of the cones on repeated use so that protein leakage contributes increasingly to the apparent free tryptophan concentration (*Flentge et al.*, 1974). However, major leakages are usually easily detectable by the visible protein precipitates on adding trichloroacetic acid in the initial step of tryptophan determination. A more serious problem derives from the lack of pH control. Plasma or serum rapidly loses CO_2 on exposure and as the pH rises the binding of tryptophan to albumin increases

(*Madras et al.*, 1974). Loss of CO_2 is not prevented by storing plasma in the frozen state. Indeed, freezing may increase losses through degassing.

When all the plasmas collected in a particular investigation are processed together (as in many animal studies) the pH changes of all specimens are usually comparable so that while absolute free tryptophan concentrations are much below their in vivo values, relative group differences are maintained and conclusions are qualitatively valid. This is less likely in clinical work when specimens may be collected and stored frozen for different periods before determination. In these circumstances, ultrafiltrate pH can be measured and free tryptophan values corrected to pH 7.5 using a standard curve obtained by determining free tryptophan of a bulked plasma brought to different pH values with CO_2 (*Sullivan et al.*, 1978). However, evidence suggests that prolonged storage at $pH > 8$ irreversibly increases plasma free tryptophan (*Kantamaneni* and *Curzon*, unpublished). Therefore it is advantageous to ultrafilter plasmas immediately on collection or (if this is not possible) to freeze under 5 % CO_2 and store tightly stoppered.

When ultrafiltration under gas pressure is used then pH can be controlled by CO_2-containing mixtures (*Bloxam et al.*, 1974; *Eccleston*, 1973) but the apparatus required is more complex and often not easily adaptable to large batches or small volumes.

A method of separating free from bound tryptophan by ultrafiltration which does not suffer from the above errors or disadvantages has been described by *Bloxam et al.* (1977). The apparatus permits ultrafiltration of small plasma samples (100—200 μl) with effective pH control and (if so desired) constant temperature—though the latter is less important. Large numbers of samples can be ultrafiltered together using units easily made from readily available material at a cost of about $ 1.— each. Except for the ultrafiltration membranes the units are re-usable.

Minor differences between findings from different laboratories may result from use of different anticoagulants (*Baumann* and *Perey*, 1977).

Non-Esterified Fatty Acid

Plasma NEFA influences tryptophan binding (*Gentil et al.*, 1977; *McMenamy*, 1965) and is therefore often determined in tryptophan studies. Numerous methods for plasma NEFA have been described, mostly based on extraction into organic solvents and either titration (*Dole*, 1956) or formation of metal soaps and subsequent metal determination (e.g. *Laurell* and *Tibbling*, 1967). Titration methods are accurate but laborious and usually require large volumes of plasma. Metal soap methods are more rapid and require little plasma but suffer from many sources of error (*Elphick*, 1968; *Williamson* and *Scott-Finnigan*, 1974). A fluorometric method (*Curzon* and *Kantamaneni*, 1977) has been developed which (like titration methods) depends on the concentration of NEFA carboxyl groups but is as rapid as metal soap methods though larger volumes of plasma are required.

Artefactual changes due to in vitro lipolysis can lead to falsely high plasma NEFA concentration unless rigid precautions are taken *e.g.* as described by *Hutson et al.* (1976). This is especially important when experimental subjects are fed on fatty diets so that plasmas are lipaemic. Thus NEFA concentrations of six lipaemic plasmas which ranged from 0.34 to 0.52 mEq/l rose after 30 min at 37 °C by amounts ranging from 0.00 to 0.48 mEq/l. The magnitude of such changes presumably depends on many factors, *e.g.* plasma lipase activity, amount and degree of aggregation of lipid substrate. In vitro increase of NEFA is not invariably apparent (*Madras et al.*, 1974) but when it occurs considerable in vitro increases of plasma free tryptophan may ensue (*Hutson et al.*, 1976) and—if unrecognized—confuse the interpretation of plasma—brain tryptophan relationships.

References

Baumann, P., Perey, M.: The analysis of free tryptophan in human blood with the ultrafiltrator: a comparison with other methods. Clin. Chim. Acta *76*, 223—231 (1977).

Bender, D. A., Boulton, A. P., Coulson, W. F.: A simple method for the study of tryptophan binding to serum albumin by small scale equilibrium dialysis: application to animals and human studies. Biochem. Soc. Trans. *3*, 193—194 (1975).

Bloxam, D. L., Warren, W. H.: Error in the determination of tryptophan by the method of Denckla and Dewey. A revised procedure. Anal. Biochem. *60*, 621—625 (1974).

Bloxam, D. L., Warren, W. H., White, P. J.: Involvement of liver in the regulation of tryptophan availability: possible role in the response of liver and brain to starvation. Life Sci. *15*, 1443—1445 (1974).

Bloxam, D. L., Hutson, P. H., Curzon, G.: A simple apparatus for ultrafiltration of small volumes: application to the measurement of free and albumin-bound tryptophan in plasma. Anal. Biochem. *83*, 130—142 (1977).

Curzon, G.: Relationships between plasma, CSF and brain tryptophan. J. Neural Transm., Suppl. 15, pp. 81—92. Wien-New York: Springer. 1979.

Curzon, G., Kantamaneni, B. D.: Fluorometric determination of plasma unesterified fatty acid. Clin. Chim. Acta *76,* 289—292 (1977).

Denckla, W. D., Dewey, H. K.: The determination of tryptophan in plasma, liver and urine. J. Lab. Clin. Med. *69,* 160—169 (1967).

Dole, V. P.: A relation between non-esterified fatty acids in plasma and the metabolism of glucose. J. Clin. Invest. *35,* 150—154 (1956).

Eccleston, E. G.: A method for the determination of free and total acid-soluble tryptophan using an ultrafiltration technique. Clin. Chim. Acta *48,* 269—272 (1973).

Elphick, M. C.: Modified colorimetric ultramicro method for estimating NEFA in serum. J. Clin. Path. *21,* 567—570 (1968).

Flentge, F., Venema, K., Korf, J.: Automated assay of tryptophan at the nanogram level: determination of tryptophan in cerebrospinal fluid and of total and non-protein bound tryptophan in serum. Biochem. Med. *11,* 231—241 (1974).

Gentil, V., Lader, M. H., Kantamaneni, B. D., Curzon, G.: Effects of adrenaline injection on human plasma tryptophan and non-esterified fatty acids. Clin. Sci. Mol. Med. *53,* 227—232 (1977).

Green, A. R.: The effects of dietary tryptophan and its peripheral metabolism on brain 5-HT synthesis and function. Essays in Neurochemistry and Neuropharmacology *3,* 103—127 (1978).

Hutson, P. H., Knott, P. J., Curzon, G.: Control of brain tryptophan concentration in rats on a high fat diet. Nature *262,* 142—143 (1976).

Knott, P. J., Curzon, G.: Free tryptophan in plasma and brain tryptophan metabolism. Nature *239,* 452—453 (1972).

Laurell, S., Tibbling, H.: Colorimetric micro determination of free fatty acids in plasma. Clin. Chim. Acta *16,* 57—62 (1967).

Lehmann, J.: Light—a source of error in the fluorometric determination of tryptophan. Scand. J. Clin. Lab. Invest. *28,* 49—55 (1971).

Madras, B. K., Cohen, E. L., Messing, R. L., Larin, F., Munro, H. N., Wurtman, R. J.: Relevance of free tryptophan in serum to tissue tryptophan concentrations. Metabolism *23,* 1107—1116 (1974).

McMenamy, R. H.: Binding of indole analogues to human serum albumin. Effects of fatty acids. J. biol. Chem. *24,* 4235—4243 (1965).

Sullivan, P. A., Murnaghan, D., Callaghan, N., Kantamaneni, B. D., Curzon, G.: Cerebral transmitter precursors and metabolites in advanced renal disease. J. Neurol. Neurosurg. Psychiat. *41,* 581—588 (1978).

Williamson, J., Scott-Finnigan, T. J.: Phospolipid interference and other variables in analysis of non-esterified fatty acids by copper soap formation. Clin. Chim. Acta *57,* 175—179 (1974).

Wood, K., Swade, C., Harwood, J., Eccleston, E., Bishop, M., Coppen, A.: Comparison of methods for the determination of total and free tryptophan in plasma. Clin. Chim. Acta *80*, 299—303 (1977).

Wurtman, R. J., Pardridge, W. M.: Summary: Circulating tryptophan, brain tryptophan, and psychiatric disease. J. Neural Transm., Suppl. 15, pp. 227—236. Wien-New York: Springer. 1979.

Author's address: Dr. *G. Curzon*, Department of Neurochemistry, Institute of Neurology, 33 John's Mews, London, WC1N 2NS, U.K.

Journal of Neural Transmission, Suppl. 15, 227—236 (1979)

Summary: Circulating Tryptophan, Brain Tryptophan, and Psychiatric Disease*

R. J. Wurtman and **W. M. Pardridge**

Laboratory of Neuroendocrine Regulation, Massachusetts Institute of Technology, Cambridge, Massachusetts, and Department of Medicine, UCLA School of Medicine, Los Angeles, California, U.S.A.

This symposium has addressed four important questions related to circulating tryptophan and the brain. These are: (1) What controls the uptake of circulating tryptophan into the brain? (2) What compounds can be measured in human blood that provide reliable indices of brain tryptophan uptake and utilization? (3) Might the metabolism of tryptophan, or its uptake into or utilization by the brain, be altered in psychiatric diseases? and, (4) Are tryptophan supplements useful in the treatment of depression and other psychiatric disorders—either by themselves or as adjuncts to drug therapy? Our views concerning these questions are summarized below. We hope that they also represent consensus opinions.

1. Factors Affecting Tryptophan Uptake in the Brain

The rate at which plasma tryptophan is delivered to the interior of brain cells depends on the transport of the amino acid through, firstly, the brain capillary wall, *i.e.*, the blood-brain barrier (BBB), and secondly, the brain cell membrane. Since amino acid transport through the plasma membranes of nerve endings and neuronal cell bodies occurs at rates much faster than that through the BBB (*Pardridge*, 1977 a), the rate-limiting step in brain uptake of plasma tryptophan is its transport through the capillary wall.

The determinants of tryptophan transport through the BBB fall into four major categories: (i) the V_{max} (maximal transport rate) of

* Supported by grants from the National Institutes of Health (AM-14228, Wurtman) and from the National Science Foundation (BNS-78-05500; Pardridge).

the BBB transport system; (ii) the K_m (app) or apparent affinity constant of the carrier for tryptophan; this represents the K_m for tryptophan alone, and the change in this K_m caused by the competition between tryptophan and the other large neutral amino acids (LNAA) for binding to the carrier; (iii) the tryptophan concentration in the plasma; and (iv) the "f" factor, or fraction of albumin-bound tryptophan that is stripped off the plasma protein by the NAA carrier during a single pass through the brain.

The V_{max}

The maximal transport rate is primarily a function of the local capillary concentration of the NAA carrier in the BBB. With a recently-developed *in vivo* competitive ligand binding assay, it has been determined that the local capillary concentration of the tryptophan carrier may be as high as, or higher than, 2500 μM (*Pardridge*, unpublished) or more than four times that of the plasma albumin, 600 μM (equivalent to an albumin concentration of 4.1 g/100 ml). The local capillary concentration of the BBB transport system (presumably a protein) is obviously much higher than could occur in free solution and illustrates one function of a membrane, *i.e.*, to concentrate proteins within small regions.

An important area of future research is the elucidation of factors that modulate the BBB carrier (generally reflected in changes in its V_{max} for tryptophan). Recent observations suggest that the LNAA carrier may be induced severalfold in animals with hepatic encephalopathy (*James et al.*, 1978) and during the neonatal period (*Sershen* and *Lajtha*, 1976); in contrast, its V_{max} is probably decreased by half in hypothyroidism (*Daniel et al.*, 1976).

The K_m (app)

The apparent affinity constant of the BBB for tryptophan is formally defined elsewhere in this volume (chapters by *Pardridge* and by *Pratt*). The K_m (app) varies directly with the concentration of LNAA in plasma. In the absence of the LNAA, the K_m is 190 μM (*Pardridge* and *Oldendorf*, 1975); at normal plasma amino acid concentrations, the K_m (app) becomes about 700 μM (*Pardridge*, 1977 b), but may range between 400 and 1000 μM.

The Plasma Tryptophan

The amount of tryptophan available for transport into the brain is equal to all of the free amino acid plus a variable fraction of the bound tryptophan (*Madras et al.*, 1974). Therefore, the "effective" tryptophan concentration may vary between two extremes: the free vs. the total tryptophan. Since under physiological conditions the fraction of bound tryptophan that is directly transported into brain exceeds the fraction represented just by the free amino acid, the *total* plasma tryptophan, and not just the free portion, usually provides a closer estimate of the actual tryptophan concentration available to the brain (see chapter by *Fernstrom*).

The "f" Factor

The fraction of bound tryptophan available for transport into brain may vary from 0 to 1.0, depending on the relative capacity/affinity ratio of BBB binding of tryptophan vs. the capacity/affinity ratio of albumin binding of the amino acid. Based on data presented previously, we calculate that the capacity/affinity ratio for tryptophan's binding to the BBB carrier may be as high as or higher than 2500 μM/190 μM = 12.5 in the absence of competing amino acids, or 2500 μM/700 μM = 3.6 in the presence of usual plasma LNAA concentrations.

Calculation of the capacity/affinity ratio for albumin requires knowledge of its concentration in the plasma, of the number of tryptophan binding sites on each albumin molecule (n = 1 in adults, n = 0.5 in neonates, see chapter by *Hamon*), and of the affinity or dissociation constant of albumin binding of tryptophan,

$$K_{diss} = \frac{\text{(trp) free}}{\text{(trp) bound}} \times \text{(Albumin)}$$

The K_{diss} may vary from 50 μM, when plasma NEFA (non-esterified fatty acids), which modulate the albumin binding of tryptophan (*Madras et al.*, 1974), concentrations are low, to 5000 μM in experimental hepatic encephalopathy (*Pardridge*, 1977 a); normally K_{diss} varies from about 170 μM to 340 μM (*Pardridge*, 1977 b). Therefore, the capacity/affinity ratio of albumin binding tryptophan is normally about 600 μM/250 μM = 2.4, or less than that for the BBB. Given these data, it is not surprising that experimental observations indicate that a major fraction of the bound tryptophan in plasma can be stripped off the albumin by the BBB carrier (*Pardridge*, 1977 b; *Yuwiler et al.*, 1977).

Examples

Table 1 describes several conditions in which changes in plasma LNAA and NEFA might, by altering the BBB's apparent K_m and albumin's K_{diss} for tryptophan, be expected to modify tryptophan's influx from the plasma to the brain.

A purely carbohydrate meal causes insulin secretion, which results in hypoaminoacidemia (decreased app K_m) and decreased NEFA (decreased K_{diss}). The net effect may be to reduce the capacity/affinity ratio of the BBB relative to that of albumin (Table 1), thus diminishing the BBB's ability to strip protein-bound tryptophan off of albumin, and blunting the enhanced influx of tryptophan into brain that would be caused by the decrease in plasma LNAA alone. If fat is included in the carbohydrate meal, plasma NEFA levels do not fall as markedly post-prandially and may even rise; the resulting increases in albumin's K_{diss} for tryptophan facilitates the "stripping" of albumin-bound tryptophan. Since insulin released by the carbohydrate/fat meal also decreases plasma LNAA, thus increasing the BBB's affinity for tryptophan, the influx of plasma tryptophan into the brain may be very rapid after such meals. (Both meals, incidentally, raise brain tryptophan levels [*Fernstrom* and *Wurtman*, 1971; *Madras et al.*, 1974].)

Table 1. *Blood-brain barrier (BBB) vs. albumin: comparison of capacity/affinity ratios for binding of tryptophan*

Condition	BBB K_m (app)[a]	BBB Capacity[b]/Affinity	Albumin K_{diss}	Albumin Capacity[b]/Affinity	BBB[c]/Albumin
Basal	700	3.5	300	2	2
Carbohydrate	400	6	100	6	1
Carbohydrate/fat	400	6	300	2	3
Protein	900	3	100	6	0.5
Phenylketonuria	5000	0.5	100	6	0.1

[a] K_m (app), *i.e.*, apparent K_m and K_{diss}, *i.e.*, dissociation constant, represent the affinity of the BBB and albumin for tryptophan, respectively; units are μM for each.

[b] The capacity/affinity ratio is the quotient of concentration (μM) of tryptophan binding sites to affinity constant (μM); the capacity for the BBB and albumin for tryptophan binding is arbitrarily set at 2500 μM and 600 μM, respectively (see text).

[c] Ratio of the capacity/affinity quotient for the BBB relative to the same for albumin binding of tryptophan; ratios of 4, 1 and 0.1, indicate 80%, 50%, and 8% of albumin-bound tryptophan will be stripped off the plasma protein by the BBB in a single passage through brain capillaries.

In contrast, a high protein meal elevates plasma LNAA, and decreases the BBB's affinity for tryptophan. Since insulin is secreted concurrently—lowering plasma NEFA and lowering albumin's K_{diss} for tryptophan—both factors tend to impede the transport of protein-bound tryptophan into the brain. Yet, as shown in Table 1, as much as 33 % of the albumin-bound tryptophan traversing the brain capillaries will still be transported into the brain. Therefore, *total* plasma tryptophan (free plus albumin-bound), corrected for LNAA, will still provide a more accurate index of the amino acid available to brain than just the free tryptophan. (Brain tryptophan levels fail to rise, or may fall [*Fernstrom et al.*, 1973] after a high-protein meal.)

Treatments like stress and salicylate administration may, by suppressing insulin secretion, elevate plasma LNAA and raise the BBB's K_m (app) for tryptophan, but have even more pronounced effects on the binding of tryptophan to albumin. Stress enhances catecholamine release, thus inhibiting insulin secretion (*Woods* and *Porte*, 1974); this accelerates lipolysis, raises plasma NEFA, and decreases the binding of tryptophan to albumin. The probable increase in albumin's K_{diss} for tryptophan will be proportionally greater than the increase in the BBB's K_m (app) resulting from the concurrent rise in plasma LNAA. Similarly, salicylates inhibit insulin secretion in the rat (*Arnold* and *Fernstrom*, 1976), raising both plasma NEFA and LNAA; but they also compete directly with tryptophan for binding to albumin, further elevating its K_{diss} (*Smith* and *Lakatos*, 1971; *Tagliamonte et al.*, 1973). Therefore, these treatments allow the BBB to compete more effectively with albumin for tryptophan binding. In contrast, the marked hyperaminoacidemia that occurs in untreated phenylketonuria greatly elevates the app K_m of BBB transport of tryptophan; this lowers the BBB's capacity/affinity ratio, and greatly impedes its ability to compete with albumin for tryptophan. It can be seen that *both* free and albumin-bound tryptophan are always available to the brain, although the proportion of albumin-bound tryptophan that is likely to be stripped off by the BBB varies, depending on what else is present in the circulation (*e.g.*, NEFA, LNAA, drugs).

Paradoxically, the good correlation that has sometimes been observed between brain tryptophan uptake and plasma free tryptophan levels after certain experimental manipulations (*e.g.*, drug administration), probably occurs *not* because the free tryptophan is the *sole*, or even necessarily the *major*, plasma moiety transported into the brain, but because the treatment has *decreased* albumin's affinity for the amino acid to the point at which the BBB carrier is maximally able to strip the amino acid off of the albumin.

2. What Compounds Should Be Measured in Human Blood and Other Body Fluids to Provide an Index of Brain Tryptophan Uptake?

The full characterization of the plasma constituents which affect brain tryptophan uptake in any physiologic or pathologic state, or after any meal or treatment, would, as described above, require measurement of plasma free and bound tryptophan levels, plasma LNAA, and plasma albumin and NEFA concentrations. Moreover, data should also be available concerning the abundance (V_{max}) of the BBB uptake system for tryptophan and other LNAA. The availability of such data would allow the investigator to calculate the capacity/affinity ratios for tryptophan's binding to the BBB uptake system and to albumin, and then to estimate the rate at which tryptophan is entering the brain. Of course, it is not possible to measure the V_{max} of the BBB uptake system directly in humans. Single-injection, venous-sampling techniques, such as the indicator-dilution method (*Yudilevich et al.*, 1972) have been used in clinical studies (*Lassen et al.*, 1971) and would provide estimates of the kinetics of tryptophan transport through the human BBB.

As described in several of the papers presented at this symposium, controversy exists as to whether plasma *free* tryptophan or plasma *total* tryptophan concentrations are predictive of brain tryptophan levels after rats receive various treatments. There seems to be general agreement that the predictive value of either of these concentrations can be enhanced by dividing them by the sum of the competing LNAA in plasma (with or without corrections for differences in transport K_m's). Unfortunately, the decision about which index of brain tryptophan uptake the investigator chooses to use (free or total tryptophan, corrected or not for LNAA) seems often to have been based on whether or not his or her laboratory was fortunate enough to possess an amino acid autoanalyzer. This brief review cannot hope fully to resolve the controversy concerning free vs. total tryptophan, save perhaps to point out once again that under almost all circumstances, the competition for tryptophan between albumin and the BBB transport system will probably be won by the BBB.

It seems safe to suggest that when the investigator is attempting to assess changes in brain tryptophan uptake after a treatment that neither causes insulin to be secreted, contributes LNAA to the blood, causes lipolysis, nor modifies plasma albumin levels, measurements of total tryptophan alone will probably suffice. (One such treatment might be the administration of small doses of tryptophan [*Fernstrom* and *Wurtman*, 1974]. However, even this is uncertain if the amino acid is accompanied by any carbohydrate, or if its dose is sufficient to stimulate insulin secretion.) Such circumstances tend to be rare, and they cannot be assumed to exist when the effects of novel treatments, or diseases of unknown etiology (*e.g.*, depression and schizophrenia), are being examined. Thus, in general, we believe that *all* studies designed to draw conclusions about brain tryptophan uptake in such diseases or experimental conditions should include measurements of plasma LNAA (and probably of plasma free tryptophan), as well as total tryptophan levels. If free tryptophan is to be measured, it cannot be done reliably by standard ultrafiltration techniques, but must instead utilize dialysis or special filtration techniques in which pH is rigorously controlled (see articles by *Fernstrom* and *Curzon* in this volume).

The possible utility of red blood cells, granulocytes, or platelets for estimating brain tryptophan levels and uptake is discussed in several accompanying papers. However, the K_m of LNAA transport into mammalian erythrocytes is ten-fold higher than usual plasma amino acid concentrations (see articles by *Young* and *Ellory*, and by *Rosenberg*). Therefore, competition effects—which are of major importance at the BBB—would not be expected significantly to affect tryptophan uptake by red blood cells (and changes in the plasma tryptophan/Σ LNAA ratio would not be expected to correlate well with red blood cell tryptophan levels) (see *Pardridge*, 1977 a).

3. Might Tryptophan Metabolism, or Its Uptake Into or Metabolism by Brain Be Altered in Psychiatric Diseases?

Several accompanying papers examine the evidence that tryptophan metabolism, brain tryptophan uptake, or serotonin synthesis are characteristically altered in one or more psychiatric diseases. We do not believe that conclusions can be drawn at the present time about tryptophan's role in the etiology of any of these diseases; more research is needed. It is important to recognize that plasma tryptophan levels normally vary widely during the day (*Wurtman et al.*, 1968), largely as the result of food intake (*Wurtman*, 1970); hence, seemingly-characteristic alterations in blood tryptophan occurring in association with a disease that affects behaviour might just as likely reflect secondary disturbances (*i.e.*, in eating rhythms or in food choice) as primary phenomena etiologically related to the disease. It is a truism to state that adequate nutritional controls are absolutely essential in any study of amino acid metabolism—especially in diseases of behavior.

4. Is Tryptophan Useful in the Treatment of Depression or Other Psychiatric Diseases?

Here too we believe that inadequate data are available on which to base a conclusion. Tryptophan appears to be a highly useful adjunct to therapy in the hands of some investigators, and useless for others. Undoubtedly some of this variation in tryptophan's effectiveness can be explained by the lack of nutritional controls in some studies: inasmuch as brain tryptophan uptake in known to be suppressed by plasma LNAA (*Fernstrom* and *Wurtman,* 1974), it should be obvious that giving tryptophan with a high-protein meal will minimize its uptake into the brain (and its possible therapeutic efficacy), while giving it with a carbohydrate (which, acting via insulin secretion, lowers plasma LNAA) should potentiate both its uptake and its efficacy. It has been suggested that another factor contributing to the variability in clinical responses to tryptophan might be the induction of a hepatic enzyme that destroys tryptophan —tryptophan pyrrolase—after people receive very high tryptophan doses. If this is true, then it should be possible to demonstrate that plasma tryptophan levels fail to rise as high after, for example, 12 g doses as they do in patients taking 6 g. Such data apparently are unavailable. Studies on rats suggest that tryptophan pyrrolase normally has relatively little to do with controlling plasma tryptophan levels (*Ross et al.,* 1973). We believe that one very likely explanation for the observed variability in clinical responses to tryptophan derives from the fact that large doses of the amino acid interfere with the uptake of tyrosine and phenylalanine into the brain (*Wurtman et al.,* 1974), and thereby suppress the syntheses of the catecholamine neurotransmitters norepinephrine (*Gibson* and *Wurtman,* 1978) and dopamine (*Scally et al.,* 1977). Tryptophan is not a "specific" drug: by inhibiting brain tyrosine uptake and suppressing the production of catecholamine neurotransmitters, it might even *exacerbate* depression in some patients. Possibly a "cocktail" of tryptophan, tyrosine, and a carbohydrate (to elicit insulin secretion and lower plasma LNAA concentrations) might be much more effective, in depression at least, than tryptophan alone. Perhaps also depression is a chemically-heterogeneous disease, with only some depressed patients suffering from inadequate serotoninergic tone (and thus being capable of responding to tryptophan supplements). Perhaps the responsiveness of groups of patients to tryptophan (or tyrosine) might even provide a new basis for classifying the depressive disorders.

References

Arnold, M. A., Fernstrom, J. D.: Salicylate reduces serum insulin concentrations in the rat. Life Sci. *19*, 813—818 (1976).

Daniel, P. M., Love, E. R., Pratt, O. E.: Hypothyroidism and amino acid entry into brain and muscle. Lancet 2, 872 (1975).

Fernstrom, J. D., Wurtman, R. J.: Brain serotonin content: increase following ingestion of carbohydrate diet. Science *174*, 1023—1025 (1971).

Fernstrom, J. D., Larin, F., Wurtman, R. J.: Correlations between brain tryptophan and plasma neutral amino acids following food consumption. Life Sci. *13*, 517—524 (1973).

Fernstrom, J. D., Wurtman, R. J.: Nutrition and the brain. Sci. Am. *230*, 84—91 (1974).

Gibson, C. J., Wurtman, R. J.: Physiological control of brain norepinephrine synthesis by brain tyrosine concentration. Life Sci. *22*, 1399—1406 (1978).

James, J. H., Escourrour, J., Fischer, J. E.: Blood-brain neutral amino acid transport activity is increased after portocaval anastomosis. Science *200*, 1395—1397 (1978).

Lassen, N. A., Trap-Jensen, J., Alexander, S. C., Olesen, J., Paulson, O. B.: Blood-brain barrier studies in man using the double-indicator method. Am. J. Physiol. *220*, 1627—1633 (1971).

Madras, B. K., Cohen, E. L., Messing, R., Munro, H. N., Wurtman, R. J.: Relevance of free tryptophan in serum to tissue tryptophan concentrations. Metabolism *23*, 1107—1116 (1974).

Pardridge, W. M., Oldendorf, W. H.: Kinetic analysis of blood-brain barrier transport of amino acids. Biochim. Biophys. Acta *401*, 128—136 (1975).

Pardridge, W. M.: Regulation of amino acid availability to the brain. In: Nutrition and the Brain (*Wurtman, R. J., Wurtman, J. J.*, eds.), pp. 141 to 200. New York: Raven Press. 1977 a.

Pardridge, W. M.: Kinetics of competitive inhibition of neutral amino acid transport across the blood-brain barrier. J. Neurochem. *28*, 103—108 (1977 b).

Ross, D. S., Fernstrom, J. D., Wurtman, R. J.: The role of dietary protein in generating daily rhythms in rat liver tryptophan pyrrolase and tyrosine transaminase. Metabolism 22, 1175—1184 (1973).

Scally, M. C., Ulus, I., Wurtman, R. J.: Brain tyrosine level controls striatal dopamine synthesis in haloperidol-treated rats. Brain Res. *41*, 1—6 (1977).

Sershen, H., Lajtha, A.: Capillary transport of amino acids in the developing brain. Exp. Neurol. *53*, 465—474 (1976).

Smith, H. G., Lakatos, C.: Effects of acetylsalicylic acid on serum protein binding and metabolism of tryptophan in man. J. Pharm. Pharmacol. *23*, 180—189 (1971).

Tagliamonte, A., Biggio, G., Vargiu, L., Gessa, G. L.: Increase of brain tryptophan and stimulation of serotonin synthesis by salicylate. J. Neurochem. *20*, 909—912 (1973).

Woods, S. C., Porte, D.: Neural control of the endocrine pancreas. Physiol. Rev. *54*, 596—619 (1974).

Wurtman, R. J., Rose, C. M., Chou, C., Larin, F.: Daily rhythms in the concentrations of various amino acids in human plasma. New Engl. J. Med. *279*, 171—175 (1968).

Wurtman, R. J.: Diurnal rhythms in mammalian protein metabolism. In: Mammalian Protein Metabolism (*Munro, H. N.*, ed.). New York: Academic Press. 1970.

Wurtman, R. J., Larin, F., Mostafapour, S., Fernstrom, J. D.: Brain catechol synthesis: control by brain tyrosine concentration. Science *185*, 183 to 184 (1974).

Yudilevich, D. L., De Rose, N., Sepulveda, F. V.: Facilitated transport of amino acids through the blood-brain barrier of the dog studied in single capillary circulation. Brain Res. *44*, 569—578 (1972).

Yuwiler, A., Oldendorf, W. H., Geller, E., Braun, L.: Effect of albumin binding and amino acid competition of tryptophan uptake into brain. J. Neurochem. *28*, 1015—1023 (1977).

Authors' address: Prof. Dr. *R. J. Wurtman*, Laboratory of Neuroendocrine Regulation, Massachusetts Institute of Technology, Building 56, Room 245, Cambridge, MA 02139, U.S.A.

Subject Index

GPSR Compliance

The European Union's (EU) General Product Safety Regulation (GPSR) is a set of rules that requires consumer products to be safe and our obligations to ensure this.

If you have any concerns about our products, you can contact us on ProductSafety@springernature.com

In case Publisher is established outside the EU, the EU authorized representative is:

Springer Nature Customer Service Center GmbH
Europaplatz 3
69115 Heidelberg, Germany

Batch number: 10577896

Printed by Printforce, the Netherlands